Rand McNally and Company, Johnson

New Universal Moulding Book

Rand McNally and Company, Johnson

New Universal Moulding Book

ISBN/EAN: 9783743687318

Printed in Europe, USA, Canada, Australia, Japan

Cover: Foto ©berggeist007 / pixelio.de

More available books at **www.hansebooks.com**

REVISED EDITION

OF THE

NEW UNIVERSAL

Moulding Book

CONTAINING LATEST STYLES OF MOULDINGS

AND

Architectural Designs of Exterior and Interior Finish

In great variety, giving full size of Mouldings, and their exact measurement in inches on each Moulding.

RAILS, BALUSTERS AND NEWEL POSTS,

ARCHITRAVES,

Front, Interior and Store Doors,

Wood Mantels, Pew Ends, Office Counters,

SCROLL AND TURNED BALUSTRADES,

Brackets and Drapery,

ELEVATIONS OF DOOR AND WINDOW FRAMES.

Entered according to Act of Congress, in the year 1891, by Rand, McNally & Co., in the Office of the Librarian of Congress, at Washington, D. C.

CHICAGO:
PUBLISHED BY RAND, McNALLY & COMPANY,
Printers, Engravers, and Electrotypers.
1891.

INDEX.

Balusters,	88
Bay Windows,	58
Brackets,	48 to 53
Chapter on Mouldings, with Illustrations,	iv, v, vi, vii
Columns and the Orders of Architecture, with Illustrations,	viii, ix, x
Cornice Drapery,	47
Counters, Office or Bank,	60
Doors,	63 to 75
Doors, Front and Frame,	73 to 75
Doors, Interior Finish,	65 to 68
Fence,	56
Frames, Window,	76 to 79
Fronts, Store,	62
Gate,	56
Glossarial Index,	xi, xii
Mantels, Wood,	61
MOULDINGS—	
Astragal,	12 and 28
Band,	18 to 22
Base,	42
Battens,	28
Bed,	6, 7
Casings,	41, 44
Ceiling and Window Stools,	38
Crown,	1 to 5
Drop Siding, Flooring and Ship Lap,	46
Eastlake and Queen Anne Casings for Windows and Doors,	42, 43
Inside Finish,	39, 40
Interior Cornice and Bead,	33, 34
Lattice, Back Band and Transom Bar,	32
Nosings,	13, 14 and 29
O G Stops,	10
Panel and Base,	15 to 17
Pew Back Rail, Wainscoting Cap and Thresholds,	30
P G and Bead Stops,	11
Quarter Round, Half Round and Cove,	9
Rabbeted Panel and Base,	23 to 27
Return Beads,	8
Section of Window Frames,	35, 36
Sprung Cove and Bed,	6
Sunk Panel,	31
Water Table or Drip Cap,	37
Pew Ends,	59
Pickets,	56
Posts, Newel,	84 to 87
Price List of Mouldings and Stair Work,	90, 91, 92
Pulpits,	59
Rail, Outside, Balusters and Posts,	55
Rail, Stair,	80 to 83
Stair Plans and Stair Brackets,	89
Veranda Sawed Balustrade and Rail,	54
Verandas,	57

CHAPTER ON MOULDINGS.

MOULDING.—A general term applied to all the varieties of outline or contour given to the angles of the various subordinate parts and features of buildings, whether projections or cavities, such as cornices, capitals, bases, door and window jambs and heads, etc. The regular mouldings of Classical architecture are, the *fillet*, or *list;* the *astragal*, or *bead;* the *cyma reversa*, or *ogee;* the *cyma recta*, or *cyma;* the *cavetto*, or *hollow;* the *ovolo*, or *quarter-round;* the *scotia*, or *trochilus;* the *torus*, or *round;* each of these admits of some variety of form, and there is considerable difference in the manner of working them between the Greeks and Romans. They are represented on page v. The mouldings in Classical architecture are frequently enriched by being cut into leaves, eggs and tongues, or other ornaments, and sometimes the larger members have running patterns of honeysuckle or other foliage carved on them in low relief; the upper moulding of cornices is occasionally ornamented with a series of projecting lions' heads.

In middle age architecture, the diversities in the proportions and arrangements of the mouldings are very great, and it is scarcely possible to do more than point out a few of the leading and most characteristic varieties. In the Norman style the mouldings consist almost entirely of rounds and hollows, variously combined, with an admixture of splays, and a few fillets (page v., fig. C); the ogee and ovolo are seldom to be found, and the cyma recta scarcely ever; in early work very few mouldings of any kind are met with, and it is not till the style is considerably advanced that they become numerous; as they increase in number, their size is, for the most part, proportionably reduced. One of the most marked peculiarities of Norman architecture is the constant recurrence of mouldings broken into zigzag lines; it has not been very clearly ascertained at what period this kind of decoration was first introduced, but it was certainly not till some considerable time after the commencement of the style; when once adopted, it became more common than any other ornament, and it is frequently used in great profusion; it may be made to produce great variety of effect by changing the section of the mouldings and placing the zigzags in different directions (figs. A and B, page v.; fig. J, page vii.). About the same time that the zigzag appeared, other ornaments of various kinds were introduced among the mouldings, and are frequently met with in great abundance; two of the most marked are the billet, and a series of grotesque heads placed in a hollow moulding, with their tongues or beaks lapping over a large bead or torus; but of these ornaments there are many varieties, and the other kinds are incalculably diversified. (Page vii., figs. E, F, K, L.)

In the Early English style, the mouldings become lighter, and are more boldly cut than in the Norman; the varieties are not very great, and in arches, jambs of doors, windows, etc., they are very commonly so arranged that if they are circumscribed by a line drawn to touch the most prominent points of their contour it will be found to form a succession of rectangular recesses, as a, b, c, d, e; they generally consist of alternate rounds and hollows, the latter very deeply cut, and a few small fillets; sometimes also splays are used; there is considerable inequality in the sizes of the round mouldings, and the larger ones are very usually placed at such a distance apart as to admit of several smaller between them; these large rounds have frequently one or more narrow fillets worked on them, or are brought to a sharp edge in the middle, as at Haddenham, Great Haseley, etc., (figs. D, G and M, page vii.); the smaller rounds are often undercut, with a deep cavity on one side, and the round and hollow members constantly unite with each other without any parting fillet or angle. The ornaments used on mouldings in this style are not numerous, and they are almost invariably placed in the hollows; the commonest and most characteristic is that which is known by the name of the tooth-ornament, which usually consists of four small plain leaves united so as to form a pyramid, but it is sometimes worked differently, and at the west door of St. Cross Church, Hampshire, and the chancel-arch of Stone Church, Kent, is composed of small bunches of leaves; these ornaments are commonly placed close together, and several series of them are frequently introduced in the same suit of mouldings; the other enrichments consist chiefly of single leaves and flowers, or of running patterns of the foliage peculiar to the style.

CHAPTER ON MOULDINGS.

GRECIAN OVOLO.
Temple at Corinth.

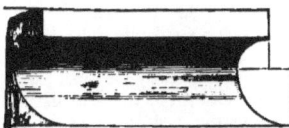

SCOTIA, TROCHILUS, OR CASEMENT.
Baths of Diocletian, Rome.

CYMA RECTA.
Theatre of Marcellus, Rome.

QUIRKED OGEE.
Arch of Constantine, Rome.

REEDS.

A
NORTH HINKSEY, BERKS.

B
IFFLEY, OXFORDSHIRE.

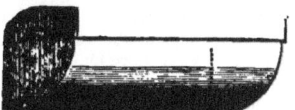

ROMAN OVOLO.
Theatre of Marcellus, Rome.

CAVETTO.
Theatre of Marcellus, Rome.

CYMA REVERSA OR OGEE.
Temple of Antonius and Faustinus, Rome.

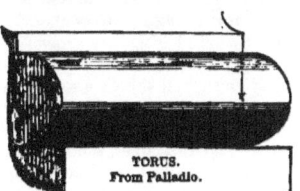

TORUS.
From Palladio.

FILLET.

ASTRAGAL.
Theatre of Marcellus, Rome.

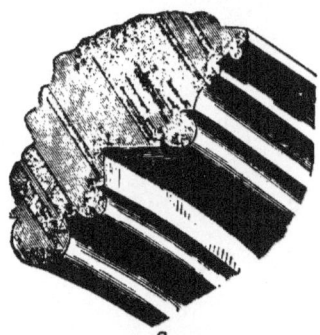

C
ARCH, CHOIR, PETERBOROUGH CATHEDRAL.

CHAPTER ON MOULDINGS.

The decorated mouldings are more diversified than the early English, though in large suits rounds and hollows continue for the most part to prevail; the hollows are often very deeply cut, but in many instances, especially towards the end of the style, they become shallower and broader; ovolos are not very uncommon, and ogees are frequent; splays also are often used, either by themselves or with other mouldings; fillets placed upon larger members are abundant, especially in the early part of the style, and a round moulding, with a sharp projecting edge on it, arising from one-half being formed from a smaller curve than the other, is frequently used; this is characteristic of decorated work, and is very common in string-courses; when used horizontally the larger curve is placed uppermost; there is also another moulding, convex in the middle and concave at each extremity, which, though sometimes found in the perpendicular style, may be considered as generally characteristic of the decorated. Fillets are very frequently used to separate other members, but the rounds and hollows often run together, as in the early English style (fig. H, page vii.). The enrichments consist of leaves and flowers, either set separately, or in running patterns, figures, heads, and animals, all of which are generally carved with greater truth than at any other period (figs. I, P, Q, R, page vii.); shields, also, and fanciful devices, are sometimes introduced; the varieties of foliage and flowers are very considerable, but there is one, the ball-flower, which belongs especially to this style, although a few examples are to be found of earlier date; this is a round hollow flower, of three petals, enclosing a ball. (Figs. N and O, page vii.)

In the perpendicular style, the mouldings are generally flatter and less effective than at an earlier period; one of the most striking characteristics is the prevalence of very large, and often shallow, hollows; these sometimes occupied so large a space as to leave but little room for any other mouldings; the hollows and round members not unfrequently unite without any line of separation, but the other members are parted either by quirks or fillets; the most prevalent moulding is the ogee, but rounds, which are often so small as to be only beads, are very abundant, and it is very usual to find two ogees in close contact, with the convex sides next each other; there is also an undulating moulding, which is common in abacuses and drip-stones, peculiar to the perpendicular style, especially the latter part of it; and another, indicative of the same date, which is concave in the middle and round at each extremity, is occasionally used in door jambs, etc. In perpendicular work, small fillets are not placed upon larger members as in decorated and early English; splays also are much less frequent than in the earlier styles, but shallow hollows are used instead. The ornaments used in the mouldings are running patterns of foliage and flowers; detached leaves, flowers, and bunches of foliage; heads, animals and figures, usually grotesque; shields, and various heraldic and fanciful devices; the large hollow mouldings, when used in arches or the jambs of doors and windows, sometimes contain statues with canopies over them.

In Normandy and the adjacent parts of France, as late as to the end of the decorated style, the mouldings do not differ materially from those of England, although there is often less variety in large suits, the same members being many times repeated; it is also very usual when capitals and bases are applied to the round mouldings in the jambs of doors and windows, etc., so as to convert them into shafts, to find that no change is made in their forms above the capitals, while, in England, the mouldings above and below the capitals are seldom the same.

CHAPTER ON MOULDINGS.

J—BEAULIEU, Near Caen, Normandy.

K—ST. EBBE'S, OXFORD. (Beak Head.)

L—ST. WILLIAM'S CHAPEL, YORK. (Chain.)

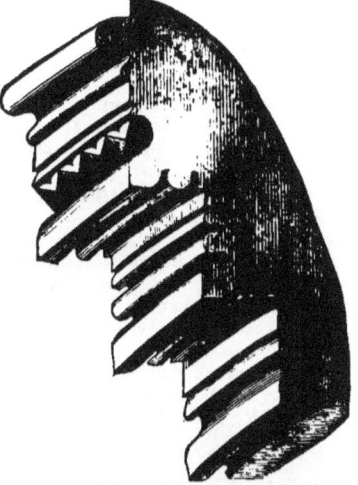

D—WEST DOOR, HASELEY, OXON. Circa 1220.

E—ABBAYE AUX DAMES, CAEN. (Segmental Billet.)

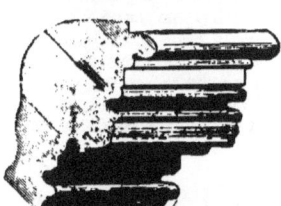

M—SOUTH DOOR ROLVENDEN CHURCH, KENT. Circa 1250.

F—ST. CROSS, HANTS. (Bird's Head.)

N—DOOR BLOXHAM, OXON. Circa 1280.

G—DOOR, PAUL'S CRAY, KENT. Circa 1230.

O—STRINGCOURSE, KIDDINGTON, OXON. Circa 1350.

P—ST. ALBAN'S HERTFORDSHIRE. Circa 1460.

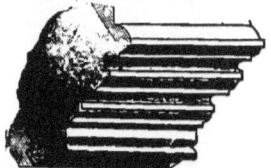

H—MERTON COLLEGE CHAPEL. A.D. 1277.

Q—ST. FRIDESWIDE'S SHRINE, OXFORD CATHEDRAL. Circa 1480.

I—SOUTHWELL MINSTER, NOTTS. Circa 1290.

R—WEST END OF NAVE, ST. MARY'S, OXFORD. A.D. 1488.

COLUMNS AND THE ORDERS OF ARCHITECTURE.

The different orders of columns in Classical architecture—(and they have been the models for those of all other styles)—are five in number: The Doric (fig. A, page ix.), Ionic and Corinthian (Figs. B and C, page ix.), and the Tuscan and Composite (figs. D and E, page ix.), which are only varieties of the Doric and Corinthian. These columns are so important an architectural feature that the exact proportions of their several parts are settled, and vary according to the order.

The *Doric* is the oldest and simplest of the three Greek Orders. Its flutings are not separated by a fillet, but by a sharp edge. The moulding below the abacus of the capital is an ovolo, but little curved in section, though quirked on the top. Below the ovolo are a few plain fillets.

The Grecian Doric has no base, or rather, all the columns stand on one base; but the Roman Doric has a separate base for each column, consisting of a plinth, torus and astragal. The ovolo is in section a full quarter-circle and is not quirked.

The distinguishing feature of the Ionic Order is its four spiral projections, called volutes. One tradition is that they are an imitation of the curls of an Ionian maiden; another that they simulate ram's horns, and still another that they are an imitation of a piece of bark placed by a builder between the echinus and the abacus, and which curled upon drying into this pleasing shape.

The principal moulding of this Order is also an ovolo, though very nicely curved.

The flutes are separated by small fillets and the lower base mouldings consist sometimes of two scotiæ, separated by small fillets and beads, above which is a large torus.

The Corinthian Order is the lightest and most ornamental of the Grecian Orders. It is said that Callimachus, a Corinthian sculptor, on observing some acanthus leaves, which had grown  up around a basket left upon a grave and had bent over after reaching the top, took it as a model for a stone capital. However it is probably an imitation of older Egyptian capitals of the same kind which still exist.

The Ionic column was higher and more slender than the Doric. The Corinthian was taller and more slender than the Ionic. Its distinguishing feature is its capital, which consists of an astragal, fillet and apopbyge, and a bell and horned abacus. The abacus consists of an ovolo, fillet and casetto.

Rows of leaves encircle the bell. The base has two scotiæ between the tori, which are separated by two astragals.

The various mouldings connected with the Orders, were, in their design and execution, such as only the Greeks, the most artistic people of the world, could produce.

A—Doric Capital and Entablature.

B—Ionic Capital and Entablature.

C—Corinthian Capital and Entablature.

D—Tuscan Capital and Entablature.

E—Composite Capital and Entablature.

F—Corinthian Base and Pedestal.

COLUMNS AND THE ORDERS OF ARCHITECTURE.

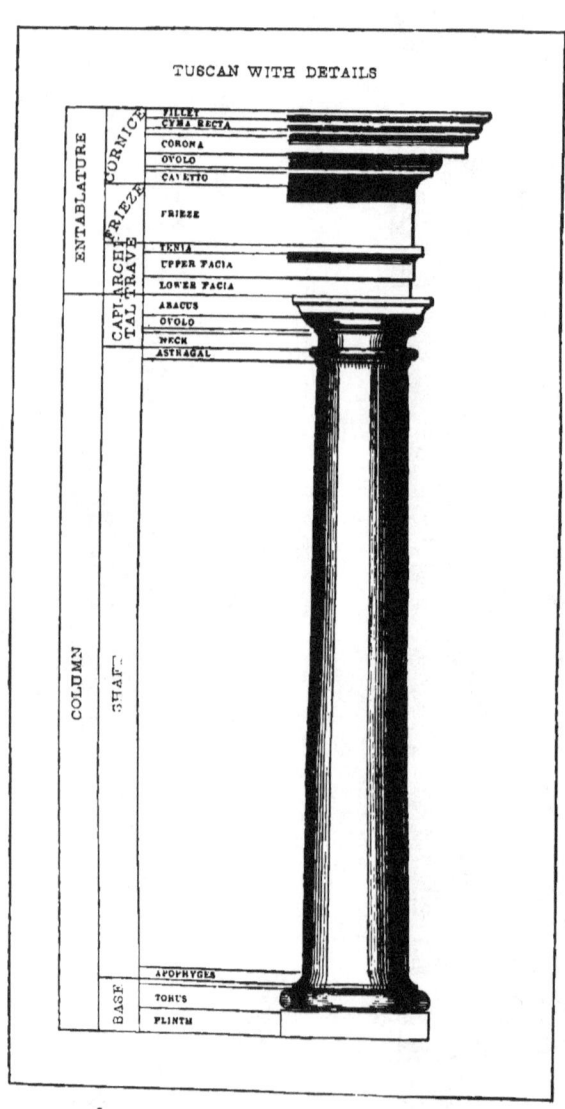

ABRIDGED GLOSSARIAL INDEX.

Abacus, the flat member at the top of a capital, originally a square tile, and in the classical styles always square: this form is retained in French Gothic, and in Norman; but in English Gothic it is usually round, sometimes octagonal.

Apse, the semicircular space at the end of a building. In Roman Basilicas the tribune. In Christian churches of the twelfth century the chancel generally terminated by an apse.

Ball-flower, an ornament peculiar to the decorated style.

Barrel-vault, resembling the inside of a barrel; called also Wagonheaded and Tunnel.

Bar-tracery, that kind of window-tracery which distinguishes Gothic work, resembling more a bar of iron twisted into various forms than stone.

Basilica, the name applied by the Romans to their public halls, either of justice, or exchange, or other business; used as churches, and afterward served as models for churches.

Battering, sloping inward from the base.

Battlement, a notched or indented parapet, originally used on castles, but afterward employed as an ornament on churches, especially in the perpendicular style.

Billet, an ornament much used in Norman work.

Byzantine Style. The term includes the styles of architecture which prevailed in the Byzantine Empire from the foundation of Constantinople, A.D. 328, to its final conquest by the Turks. The plan of these buildings was generally that of a Greek cross, with a large cupola rising from the center, and a smaller one over each of the arms of the cross, and sometimes two on the nave: arches, round or horseshoe.

Chamfer, a sloping surface forming the bevelled edge of a square pier, moulding: the two ends of the chamfer are often ornamented, and these ornaments are called chamfer-terminations.

Clere-story, or *Clear-story*, the upper story of a church, as distinguished from the triforium or blind-story below it, in which the openings, though resembling windows, are usually blank or blind, not glazed.

Clustered Pillar, a pillar formed of a cluster of small shafts, or made to appear so, and a distinguishing feature of the Gothic styles. In classical and Norman architecture the columns are plain and not clustered, and this is often the case in French Gothic also.

Coping, the sloping termination of a wall or buttress, to throw off the water; when forming the top of a buttress, it is also called a gablet, or little gable.

Corbel, a projecting stone to carry a weight, usually carved. In Norman work the corbels are often made into grotesque heads, and the eaves of the roof are carried by a row of corbels called a corbel table. In the early English style the corbels are often carved into the form called a mask or a buckle, but heads are also commonly used, or foliage. In the decorated style they are often the heads of a king and a bishop, especially those carrying the drip-stone over a door, or a window. In the perpendicular style the moulding is often continued, and forms a square or round termination, called a drip-stone termination.

Crocket, an ornament peculiar to the Gothic styles, usually resembling a leaf half opened, and projecting from the upper edge of a canopy or pyramidal covering. The term is supposed to be derived from the resemblance to a shepherd's crook. It is not used in the Norman style.

Cruciform Churches. In the western parts of the Roman Empire the Latin form was adopted, *i.e.*, the nave long, the choir and transepts short; in the eastern, the Greek form, *i.e.*, the four arms of equal length.

Crypt, a vault beneath a church, generally beneath the chancel only; used sometimes for exhibition of relics.

Decorated English Style, the second Gothic style. Windows, showing geometrical tracery.

Early English Style, the first Gothic style.

Elizabeth, the style of the Renaissance, and a mixed style.

Entablature, the horizontal block of stone or masonry, lying across the top of two columns, found in classical architecture; but by degrees the arch substituted for it.

Facia, a broad fillet, band, or face used in classical architecture, sometimes by itself, but usually in combination with mouldings.

Fillet, a small square band used on the face of mouldings.

Finial, the ornament which finishes the top of a pinnacle, a canopy, or a spire, usually carved into a bunch of foliage.

Gable, the end wall of a building sloping to a point.

Gablets, small gables.

Gargoyle, or *Gurgoyle*, a projecting water-spout, often ornamented with grotesque figures.

Geometrical Tracery. This term is applied when the openings are of the form of trefoils, quatrefoils, spherical triangles, etc. This kind of tracery came into use in the time of Edward I.

ABRIDGED GLOSSARIAL INDEX.

Gothic, the style of architecture which flourished in the western part of Europe from the end of the twelfth century to the revival of the classical styles in the sixteenth.

Jambs, the sides of a window opening, or doorway.

Lancet Window, a window the lights of which are of the form of a surgeon's lancet, chiefly used in the thirteenth century, but occasionally at all periods. At first they are single, then two, three, or more together separated by solid masonry, which is gradually reduced in thickness until mere *mullions* are produced; several lancets are then grouped under a single arch.

Mask or *Buckle*, an ornament used on corbels in the thirteenth and fourteenth centuries; when looked at in front it often resembles a buckle, but the shadow of it on the wall is the profile of a human face.

Monastery, in early times—a church with three or four priests attached, often called *Minster*.

Mullion, the vertical bar dividing the lights of a window; it occurs in very late Norman work, but is essentially a Gothic feature.

Niche, or *Tabernacle*, a recess for an image.

Norman Style, commences in the last quarter of the eleventh century, and ceases during the last quarter of the twelfth.

Oculus, a term applied to the large circular window at the west end of a church, common in foreign churches, but not usual in England.

Ogee, a moulding formed by the combination of a round and hollow.

Parapet, the low wall at the top of a building forming the outline against the sky, at first solid, then often divided into battlements, afterward pierced with ornamental open-work.

Pier-arches. The main arches of the nave or choir resting on piers are so called.

Pinnacle, a sort of small spire usually terminating a buttress.

Plinth, the projecting member forming the lower part of a base, or of a wall.

Pointed. First, middle, and third pointed styles, synonymous with the more generally received names of early English, decorated and perpendicular.

Pointed Arch. This is usually a Gothic feature, or a mark of transition to it, but it occurs also in earlier work, before the change of style, as at Fountains Abbey, Malmesbury.

Porch, a projecting structure to protect a doorway.

Quoins, corner stones.

Renaissance, Style of the, in England called Elizabethan, or Jacobean.

Rib, a projecting band or moulding on the surface of a vault.

Romanesque, the French term for the debased Roman styles, including the *Norman* style.

Spherical Triangle, a triangular opening with curved sides, used in clere-story windows, as at Cranford; and in the tracery lights of other windows, as at Morton and York.

Spire, an essentially Gothic feature.

Squinches, the small arches across the angles of a square tower to carry an octagonal spire.

Stone Churches, first built about A.D. 680.

Sunk-chamfer Moulding, a feature of the decorated style.

Tooth-ornament, an ornament resembling a row of teeth, sometimes called Dog's Tooth and Shark's Tooth. M. de Caumont and the French antiquaries call it *Violette*, and it often bears considerable resemblance to that flower when half expanded; it occurs in Anjou in work of the twelfth century; in England it is rarely used before the thirteenth, when it is so abundant as to form one of the characters of the early English style. In France it is used freely in Normandy, but scarcely at all in the Domaine Royale.

Transept, the portion of the building crossing the nave, and producing a cruciform plan.

Transition. The period of a change of style, during which there is frequently an overlapping of the styles, one building being in the old style, another in the new, at the same period. The last quarter of each century was a period of transition, or change from the style of that century to the style of the one which came after. This term is chiefly applied to the great change from the Norman, or Romanesque, to the Gothic style, but may also be applied, in a minor degree, to each of the subsequent changes of style.

Transom, the transverse horizontal piece across the mullions of a window; it occurs sometimes in early English, and decorated work, but is far more common in the perpendicular style.

Tribune, the semicircular space at one end of the basilica, for the judges. In churches copied from the basilicas, it was retained as the apse.

Triforium, or blind-story, the middle story of a large church, over the pier-arches and under the clere-story windows; it is usually ornamented by an arcade, and fills the space formed by the necessary slope of the aisle roofs.

Tudor-flower, an ornament belonging to the perpendicular style, but not confined to the Tudor period.

Turrets, small towers.

Twelfth Century. Vide *Norman*.

Tympanum, the space between the flat lintel of a doorway and the arch over it, usually filled with sculpture

CROWN MOULDINGS.

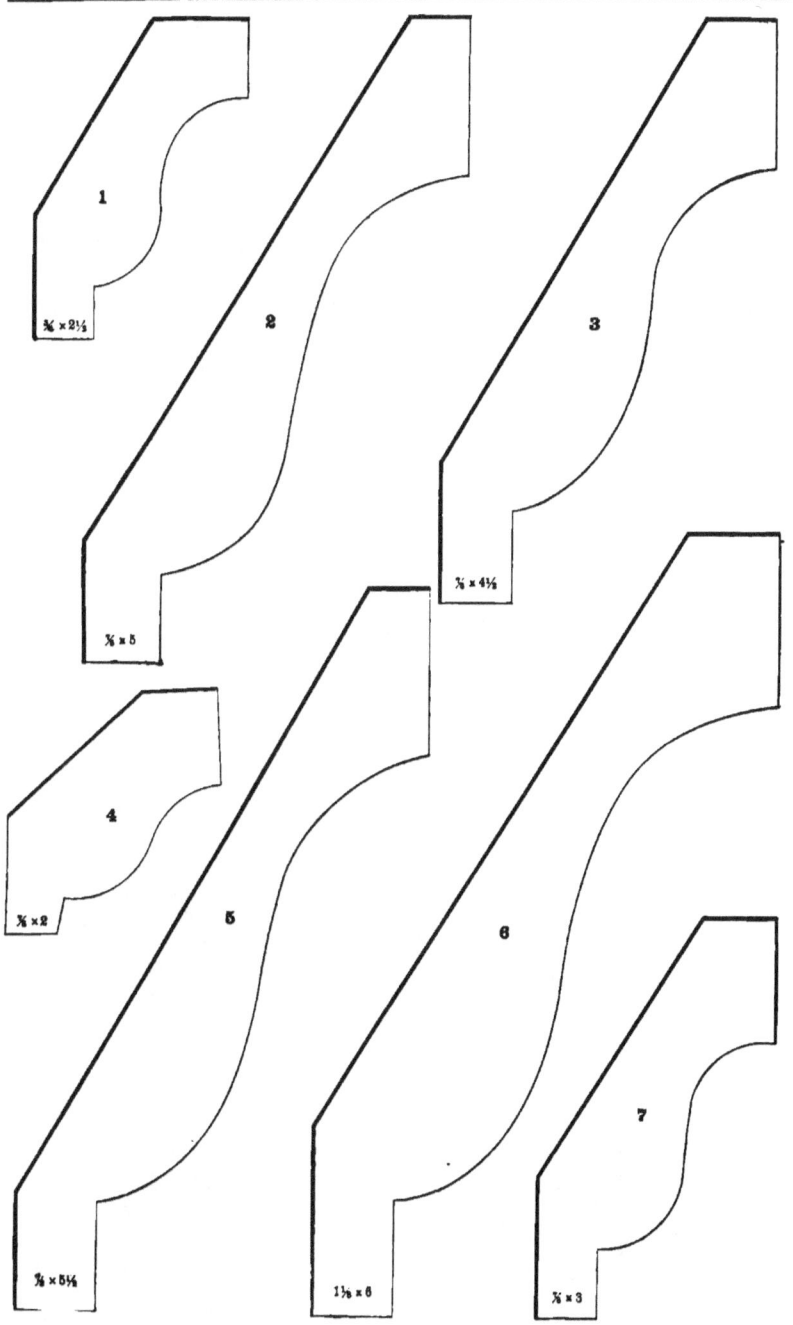

CROWN MOULDINGS.

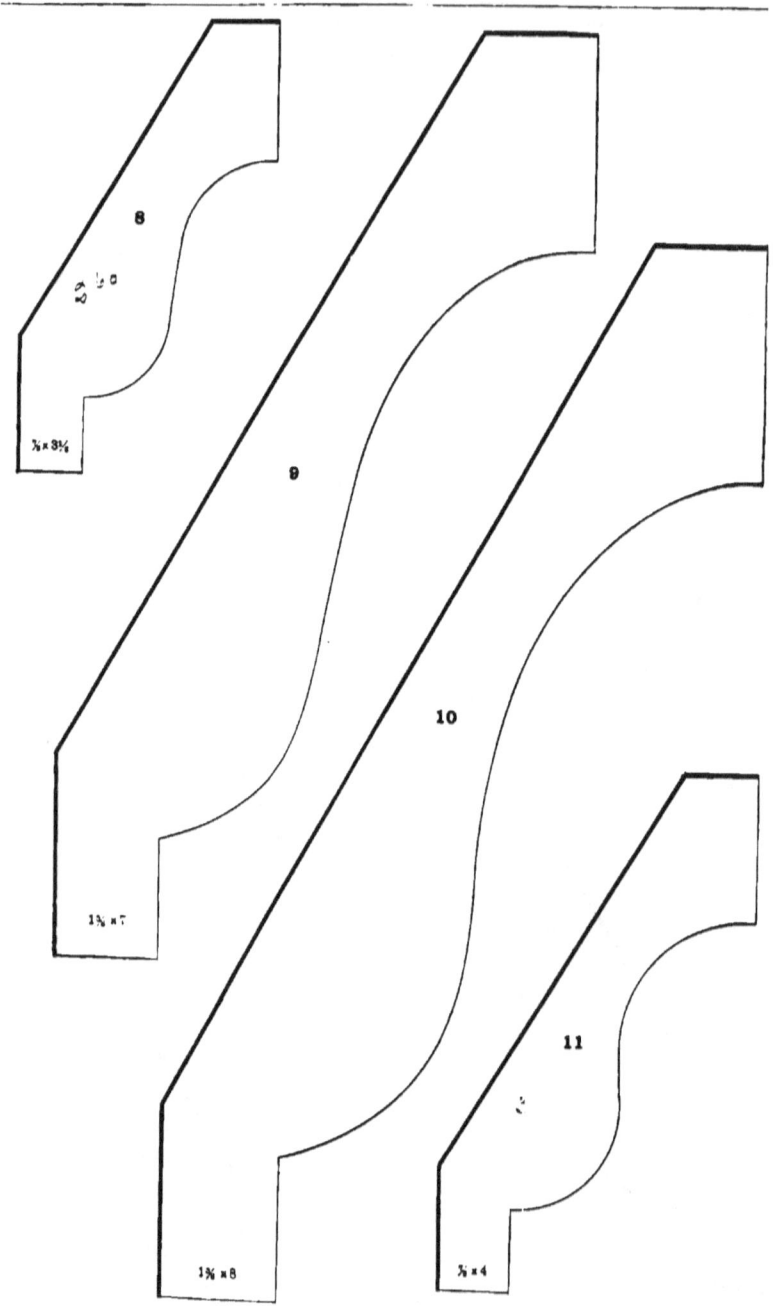

CROWN MOULDINGS.

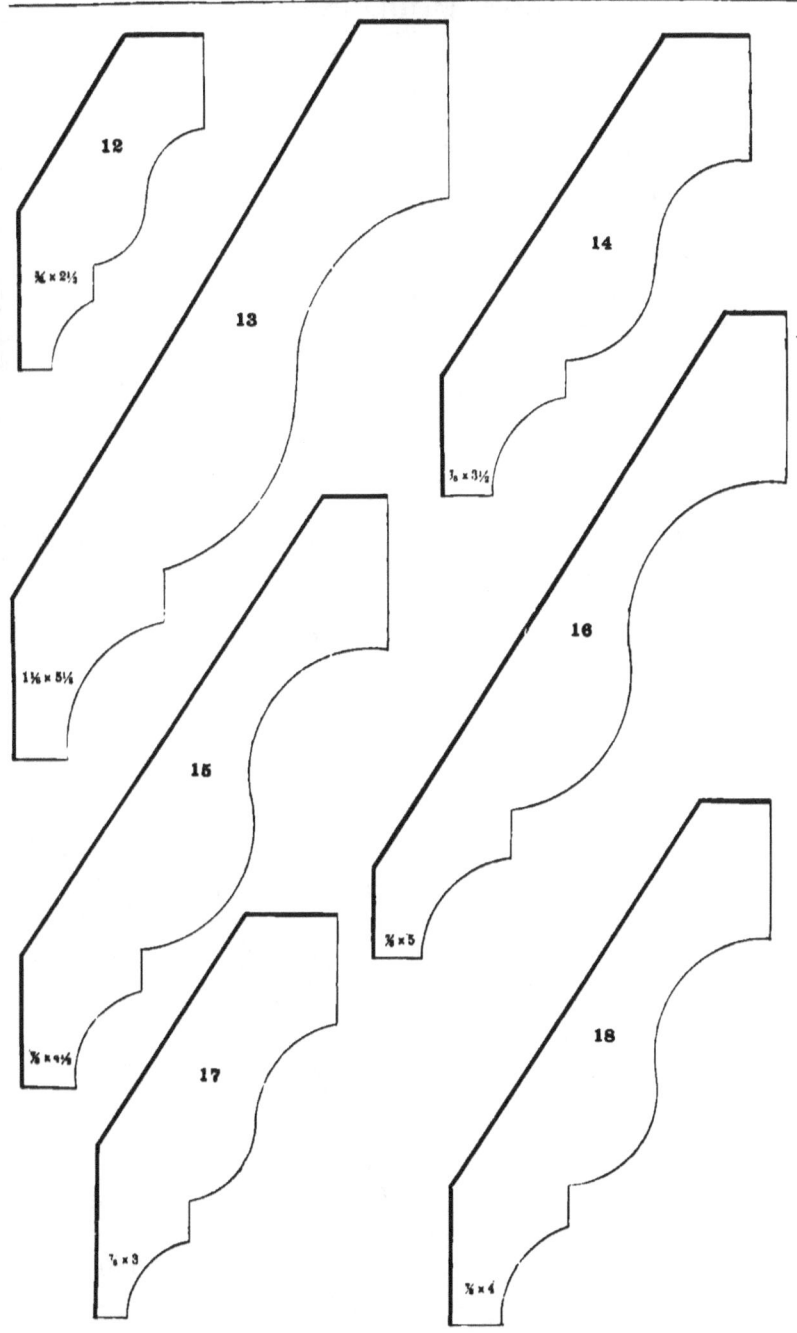

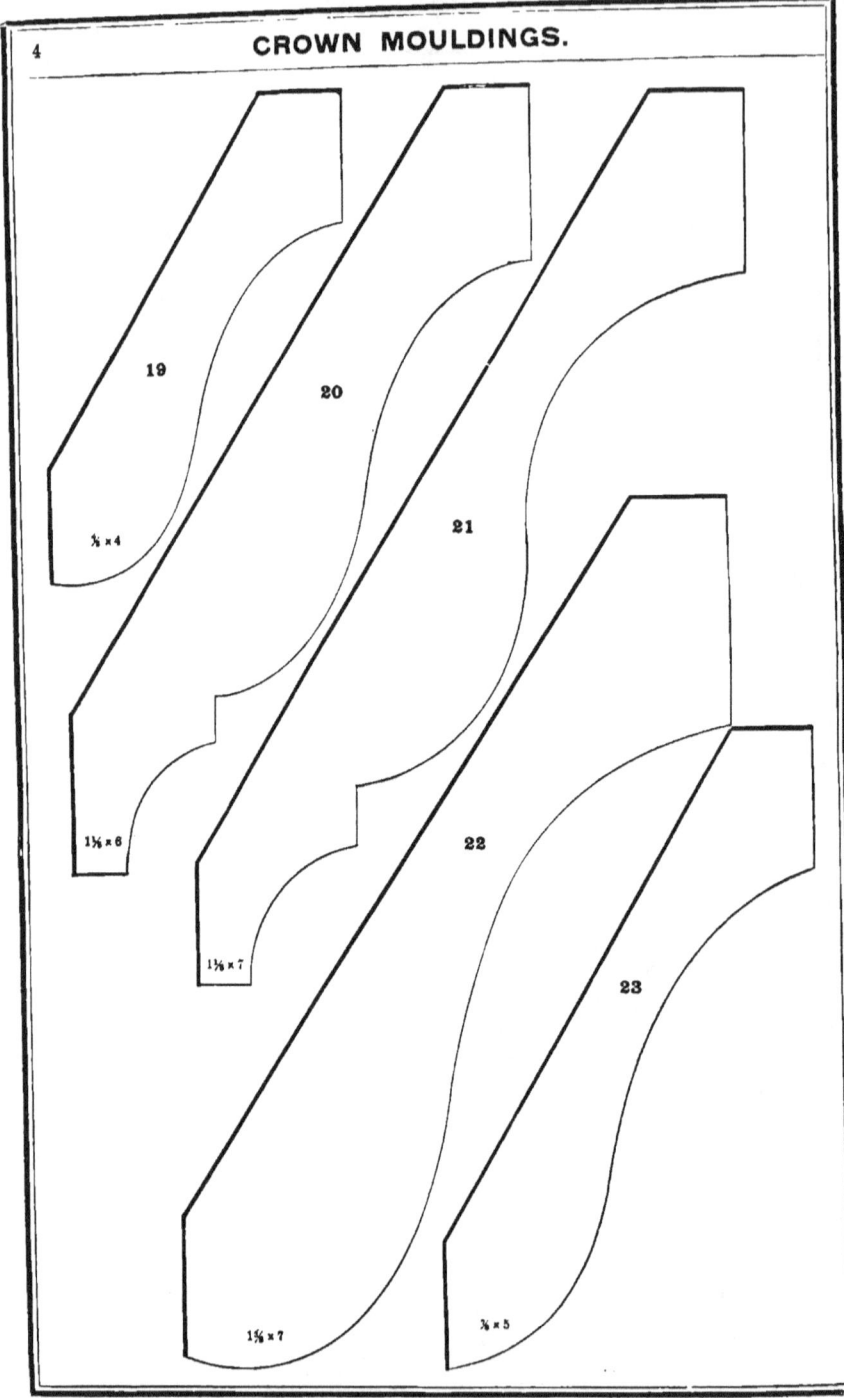

CROWN MOULDINGS.

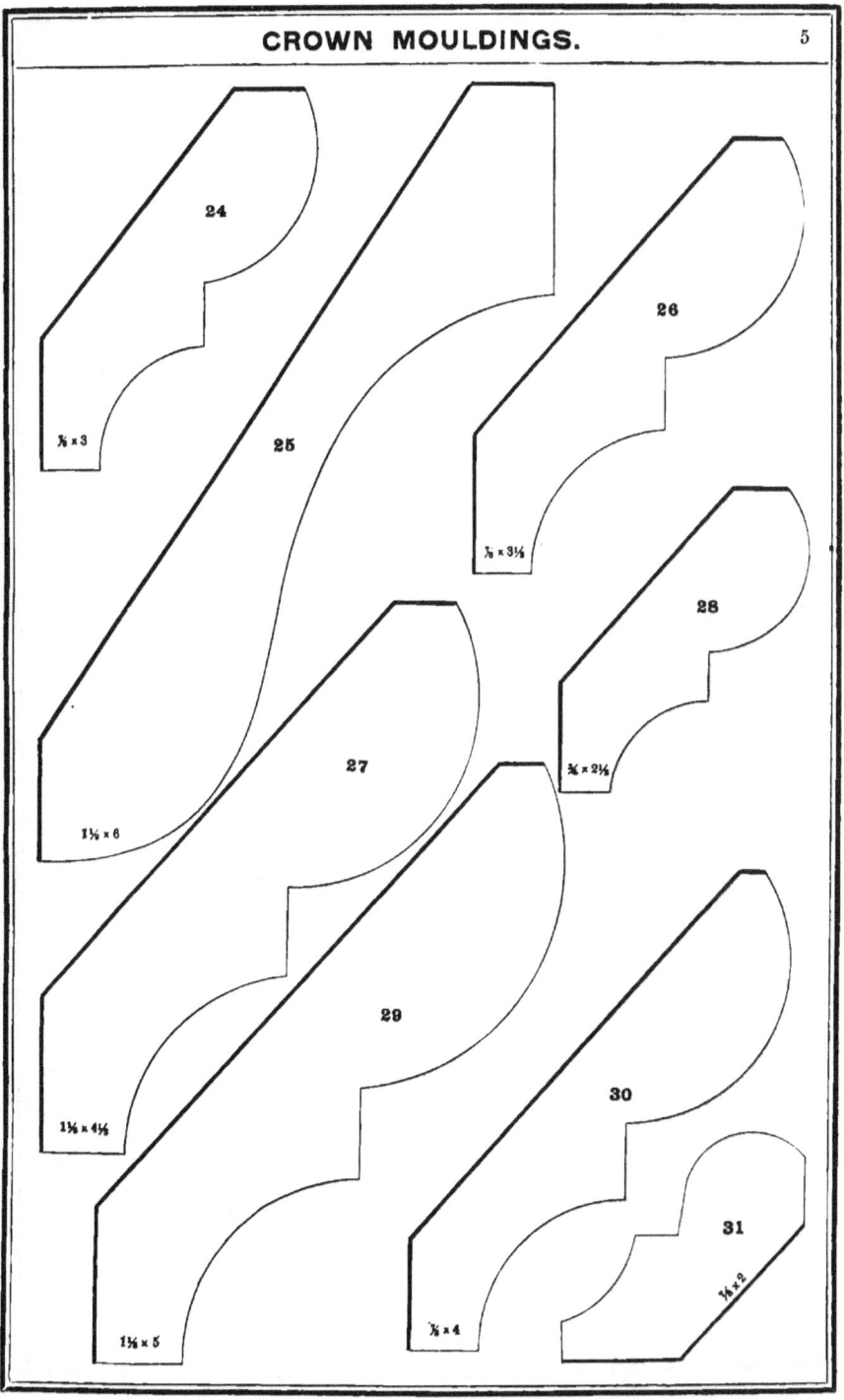

SPRUNG COVE AND BED MOULDINGS.

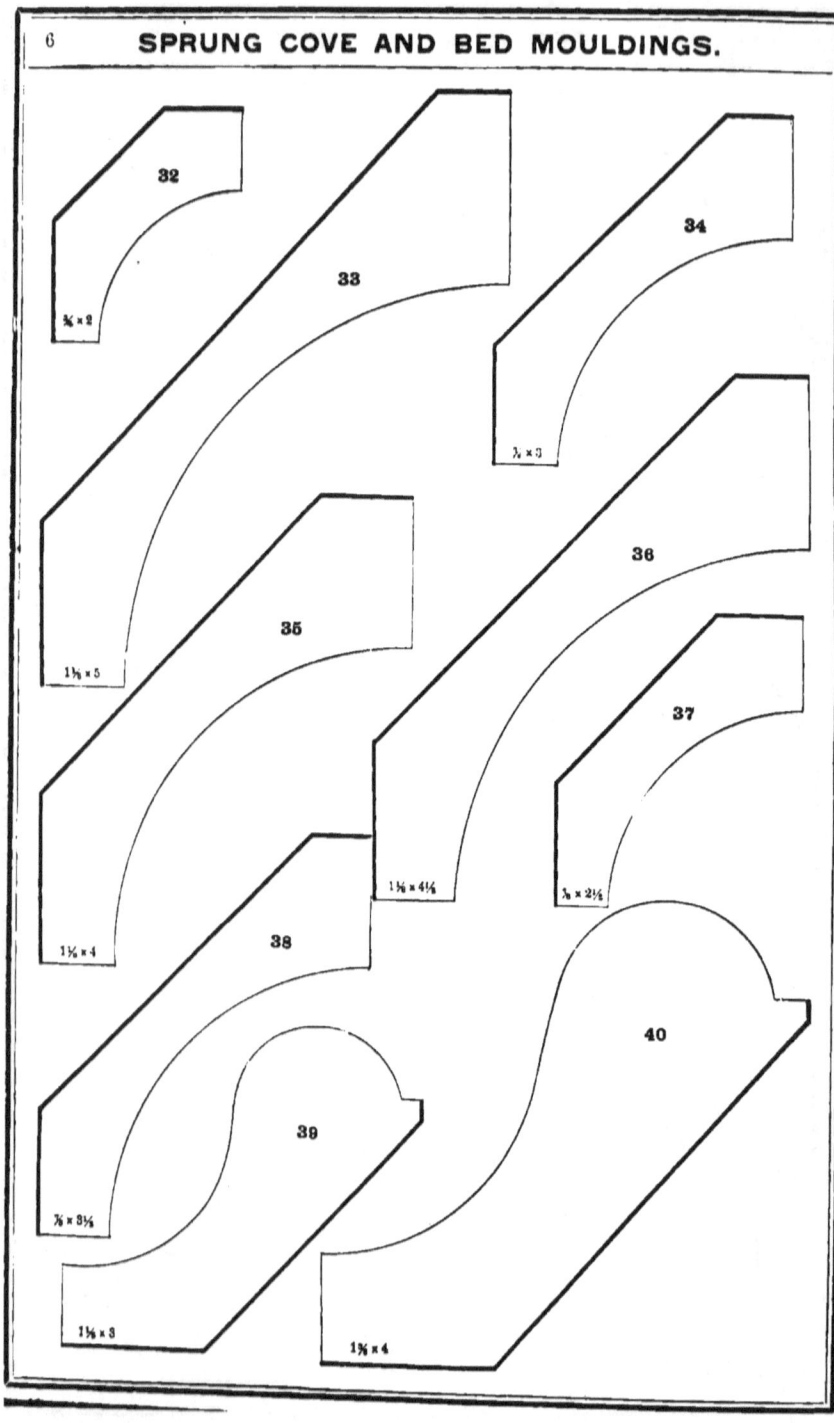

BED MOULDINGS.

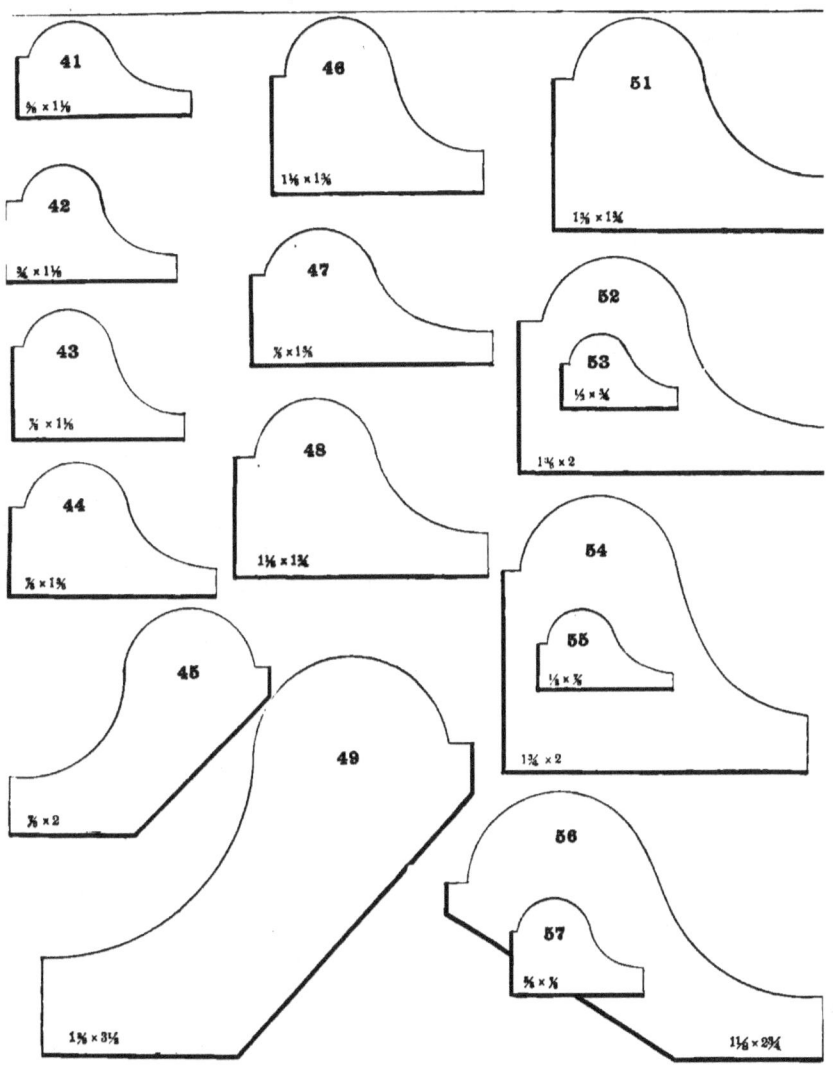

RETURN BEADS.

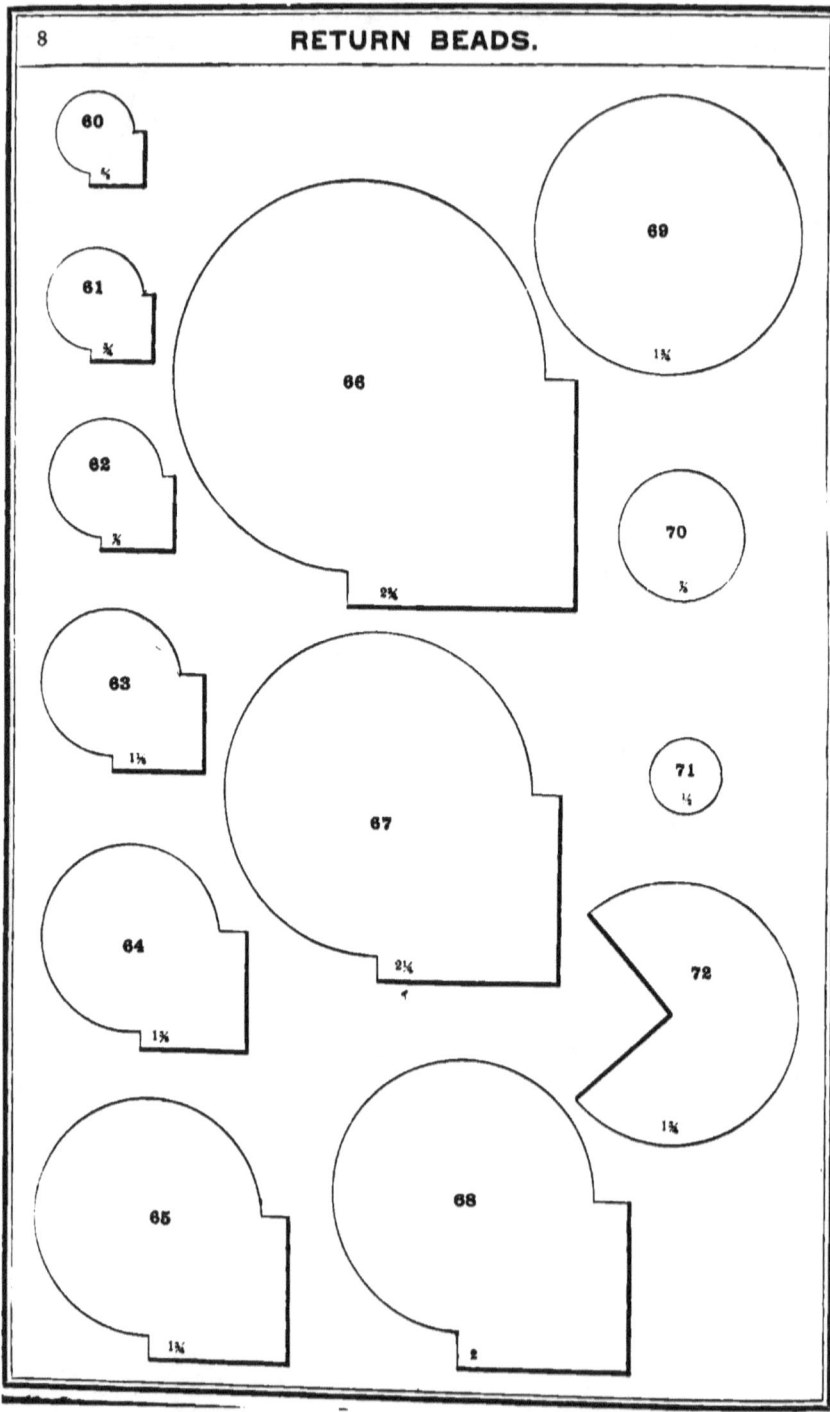

QUARTER ROUND, HALF ROUND AND COVE.

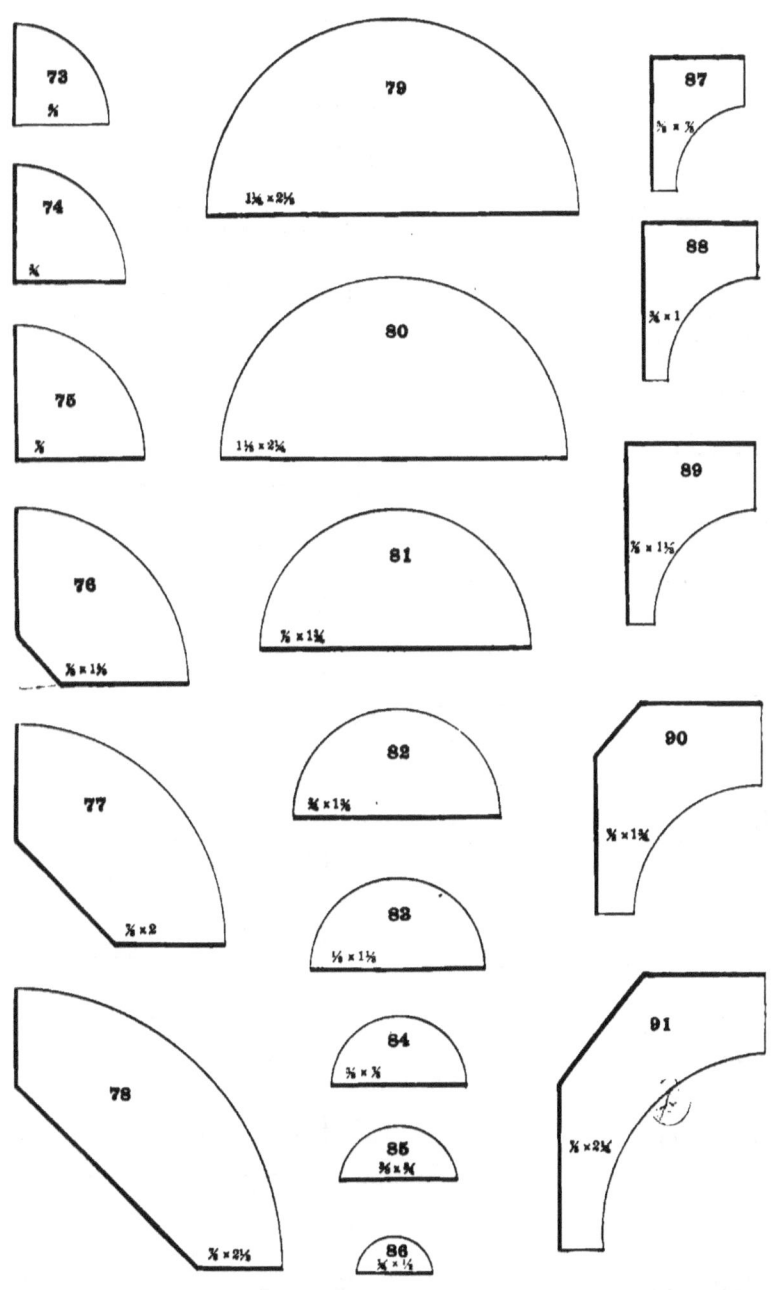

O G STOPS.

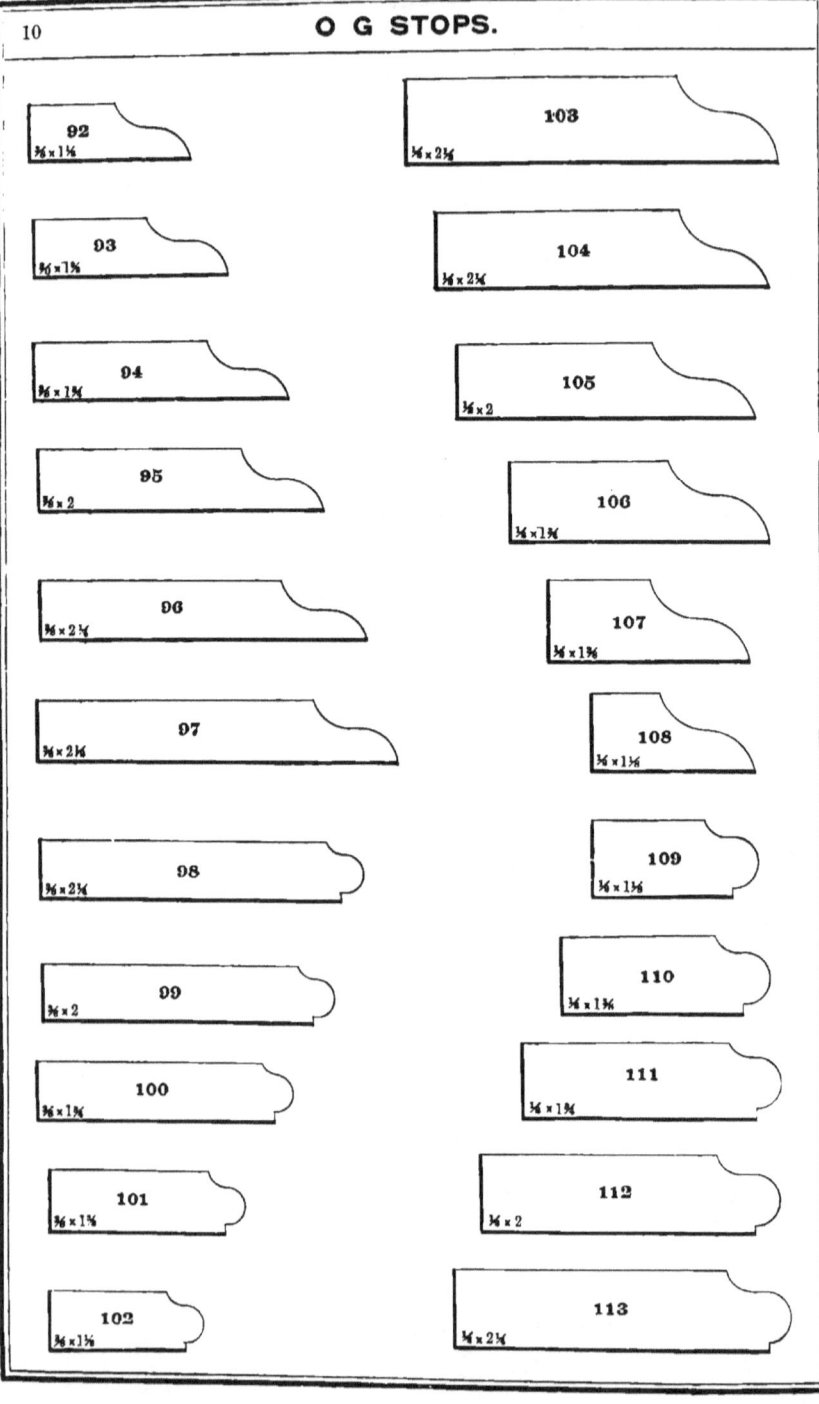

P. G. AND BEAD STOPS.

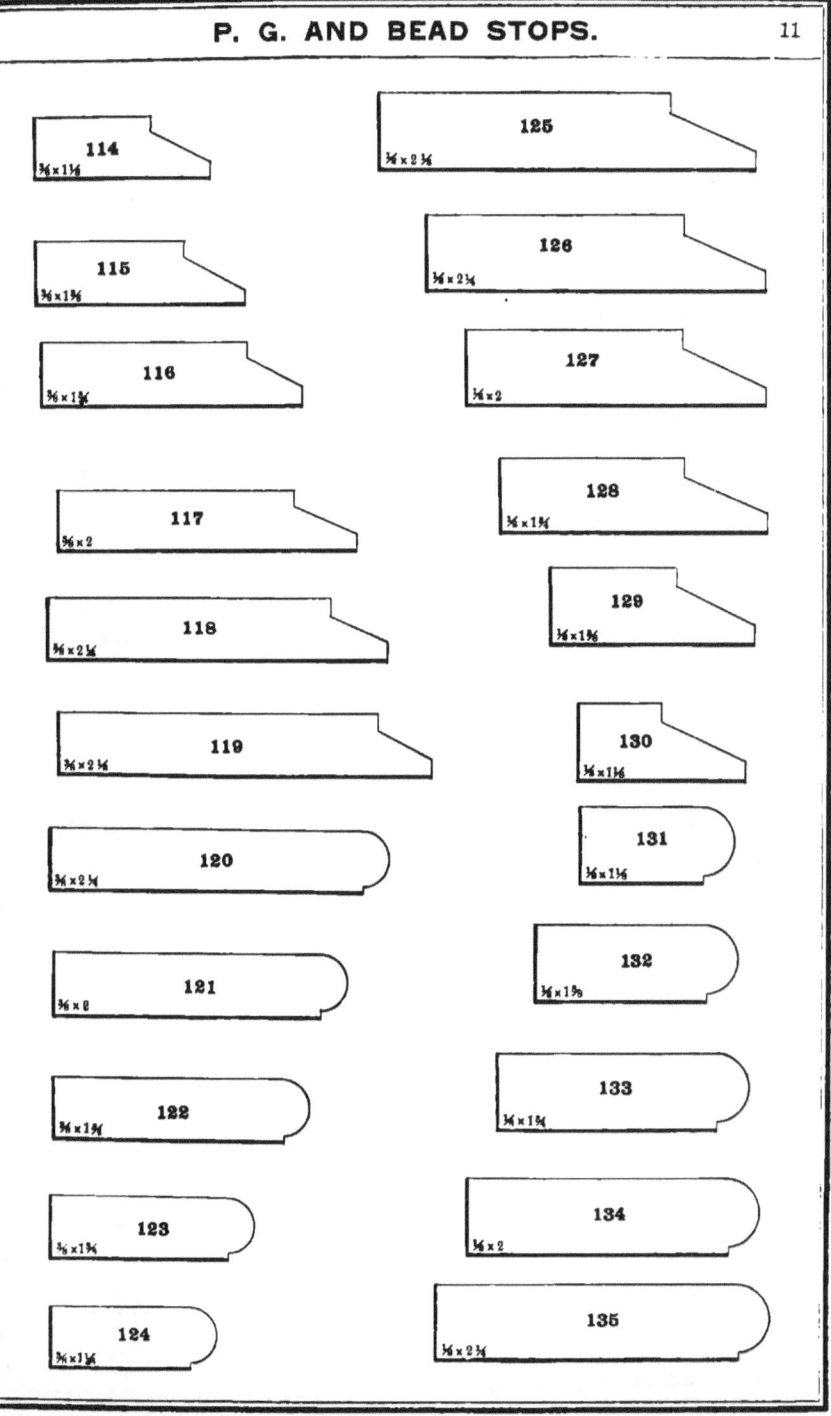

ASTRAGAL MOULDINGS.

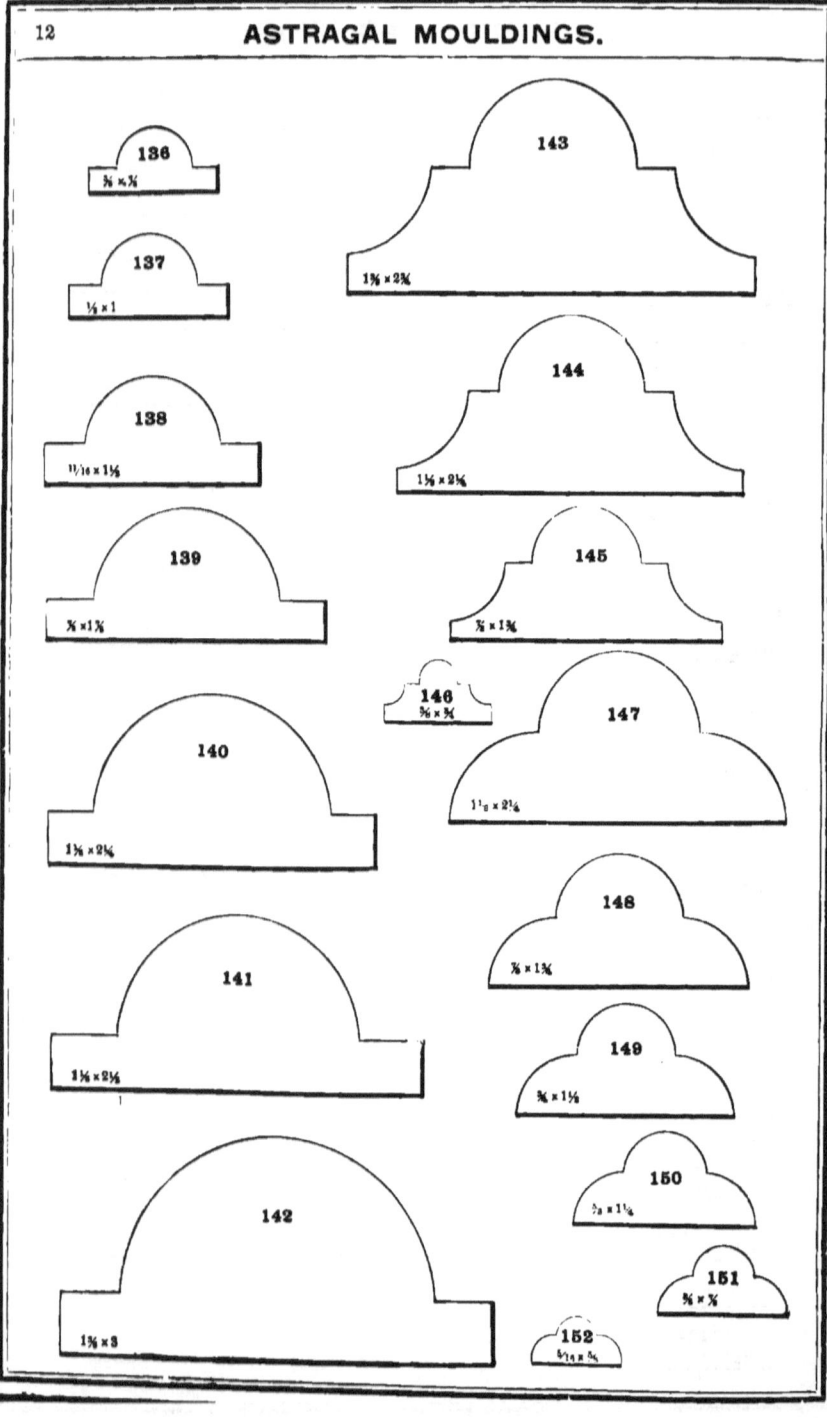

NOSINGS.

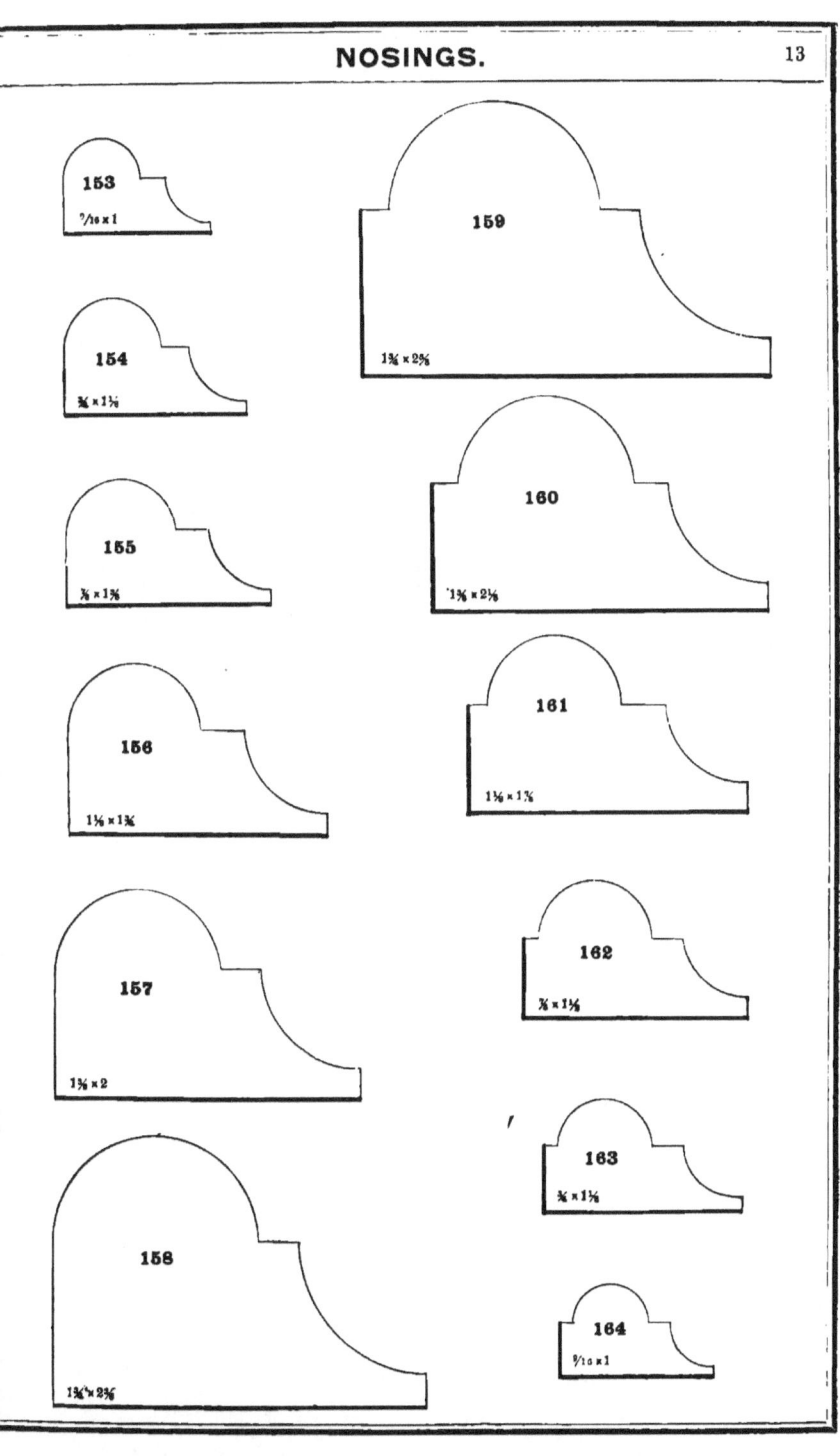

NOSINGS.

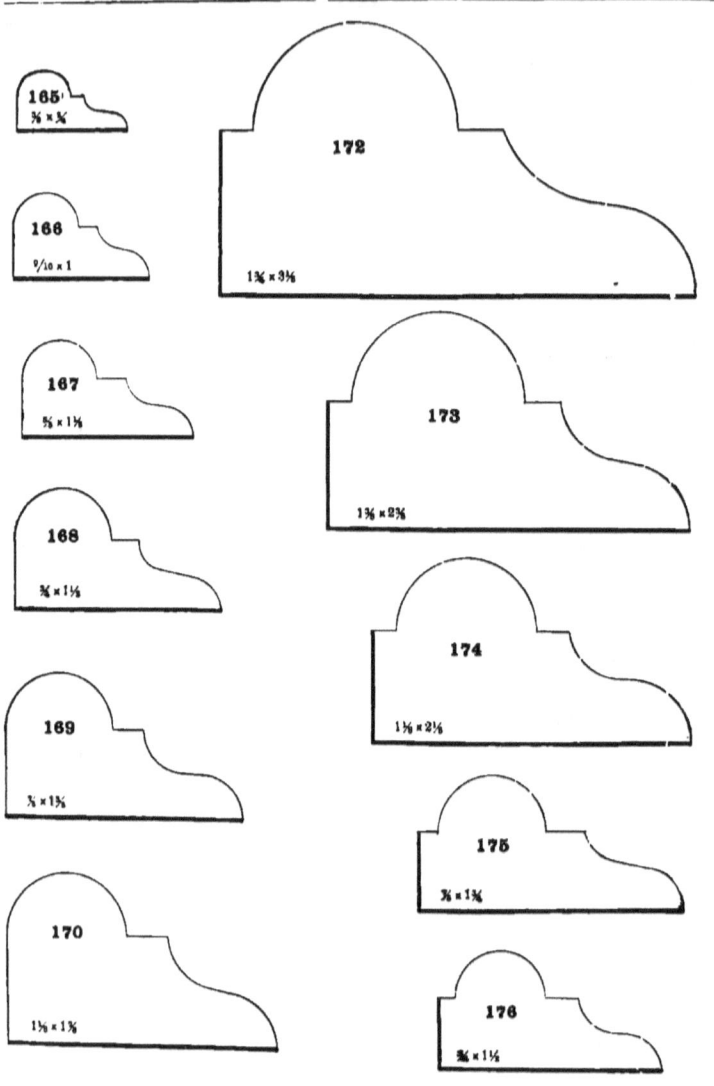

PANEL AND BASE MOULDINGS. 15

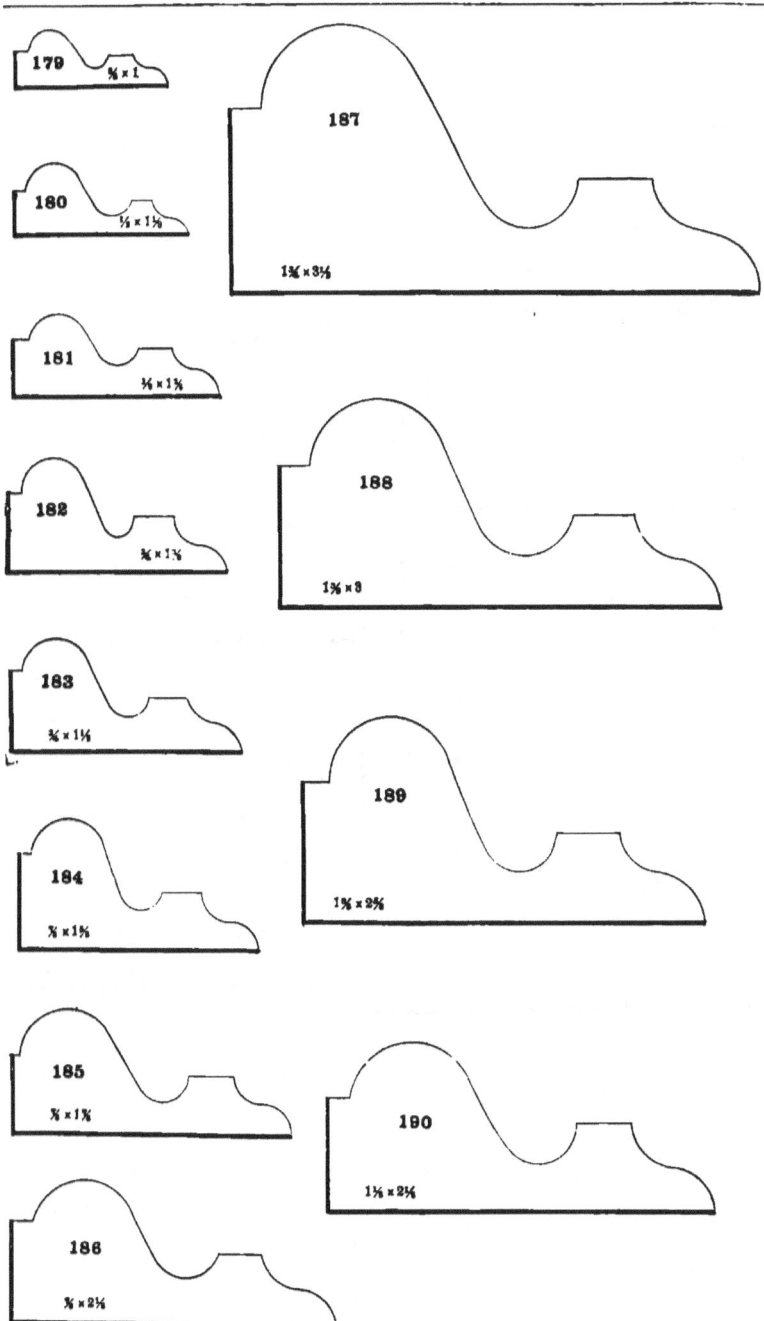

PANEL AND BASE MOULDINGS.

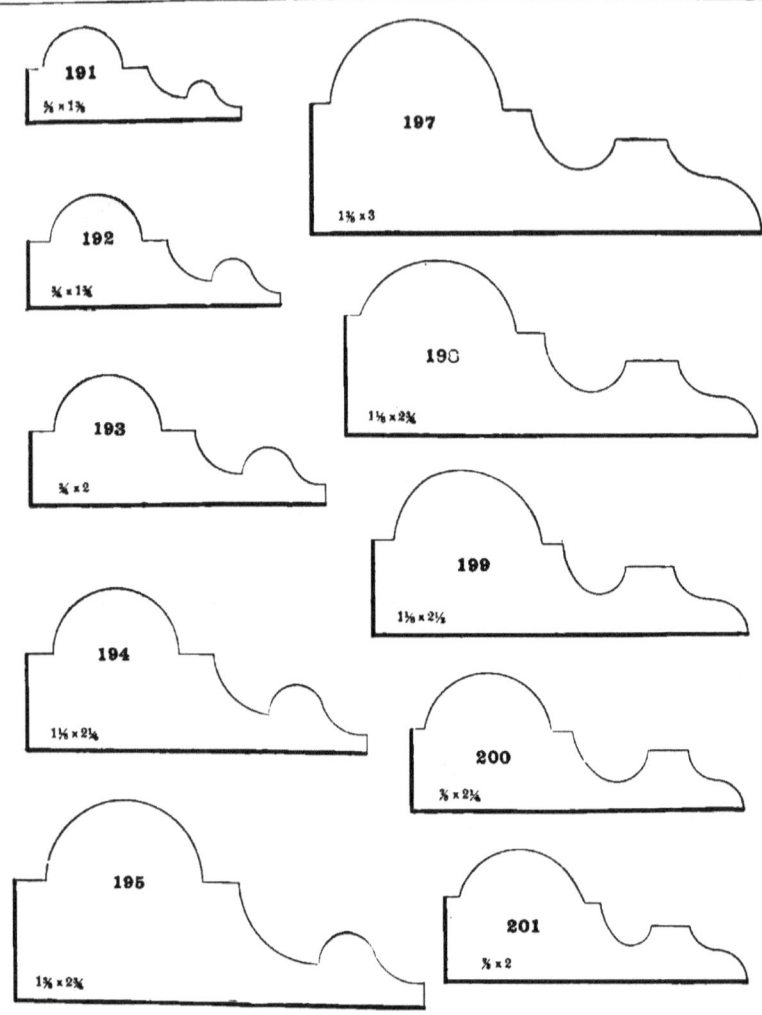

PANEL AND BASE MOULDINGS.

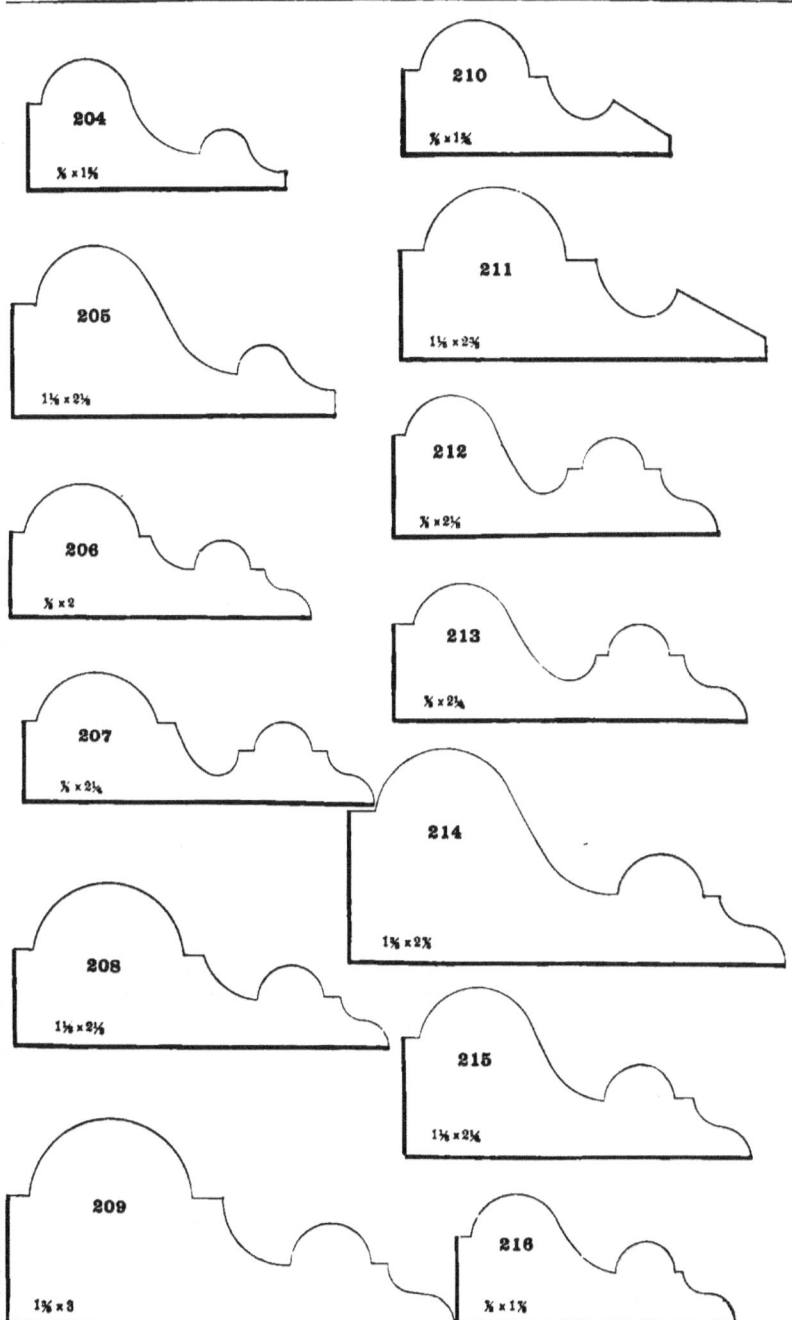

BAND MOULDINGS.

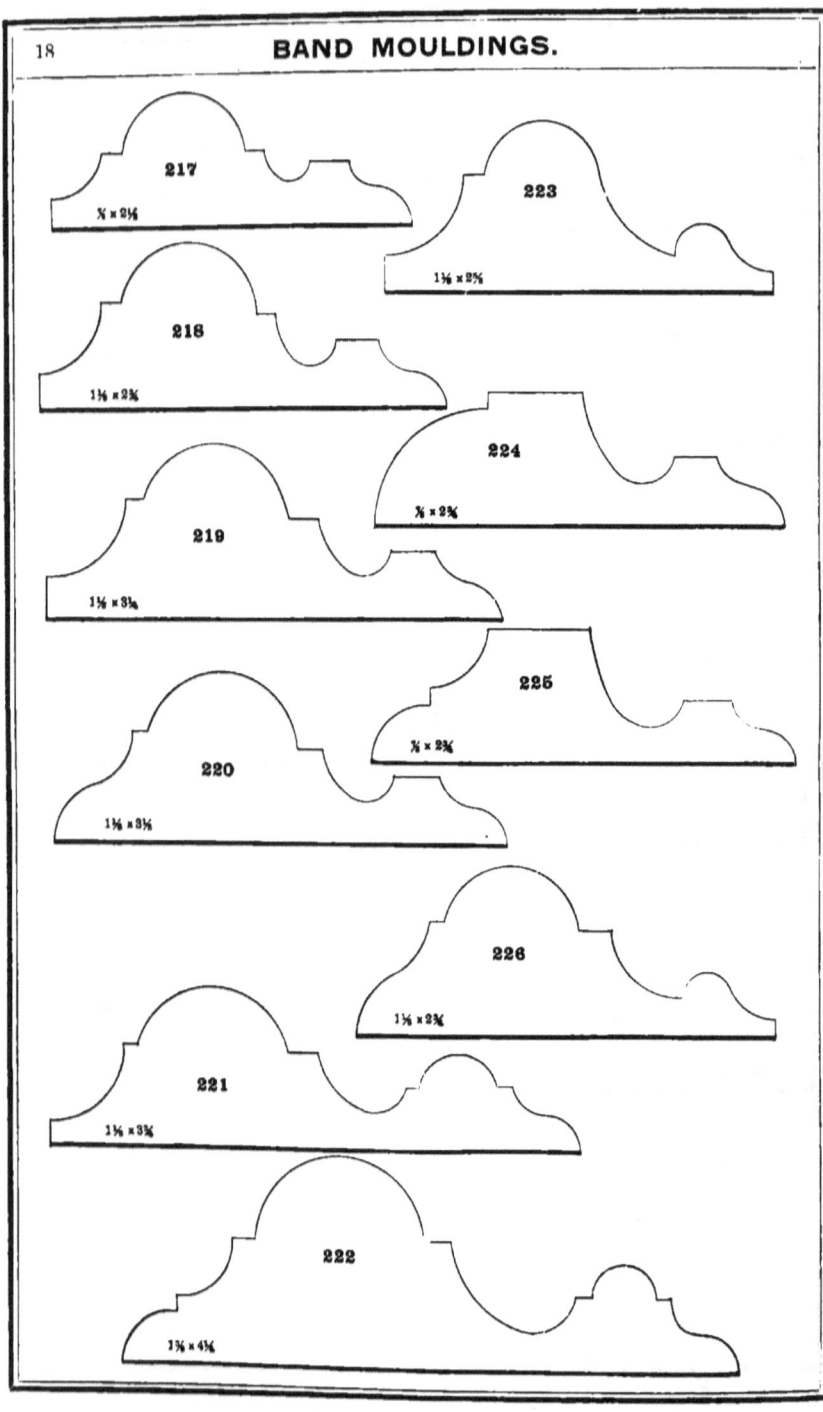

BAND MOULDINGS.

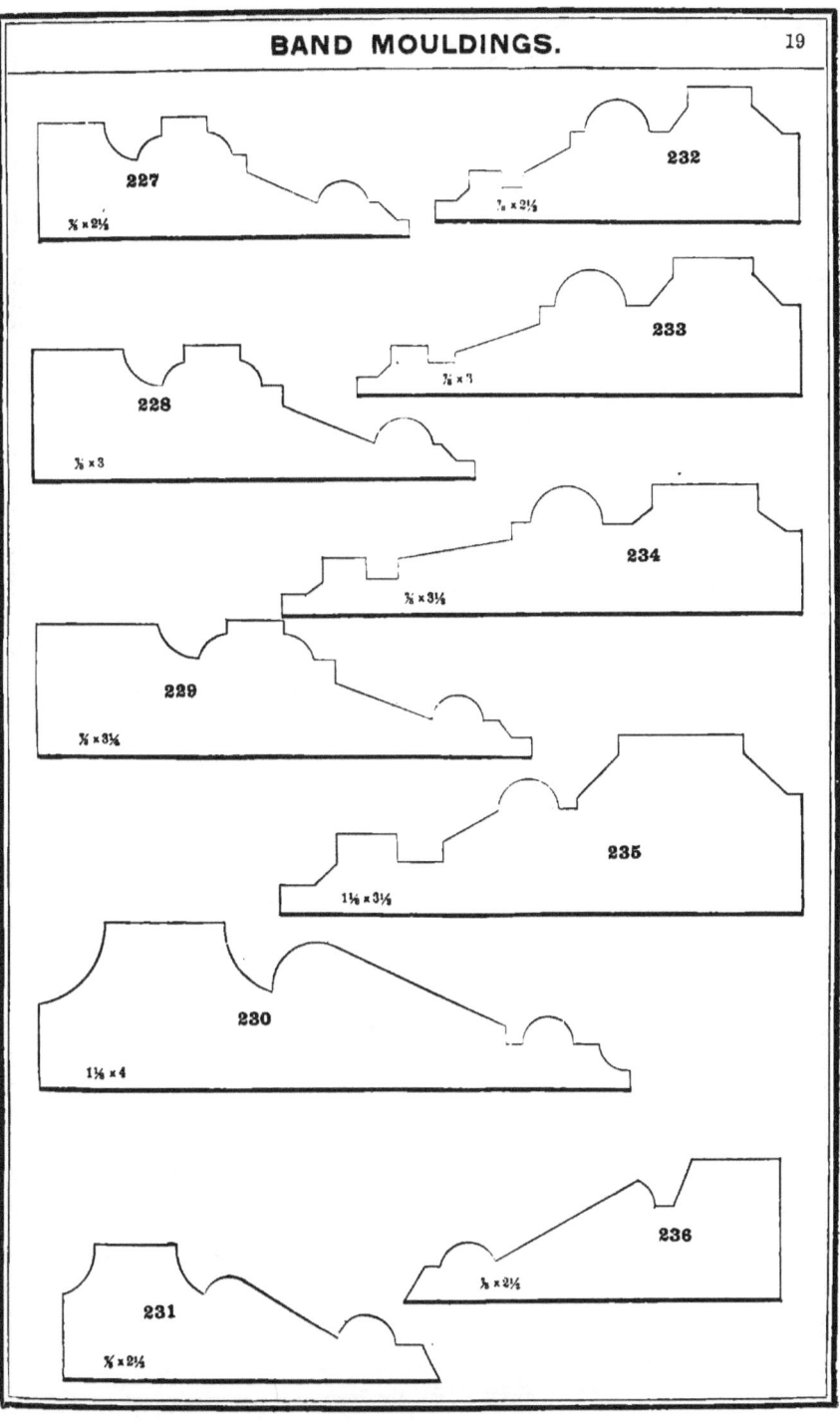

BAND MOULDINGS.

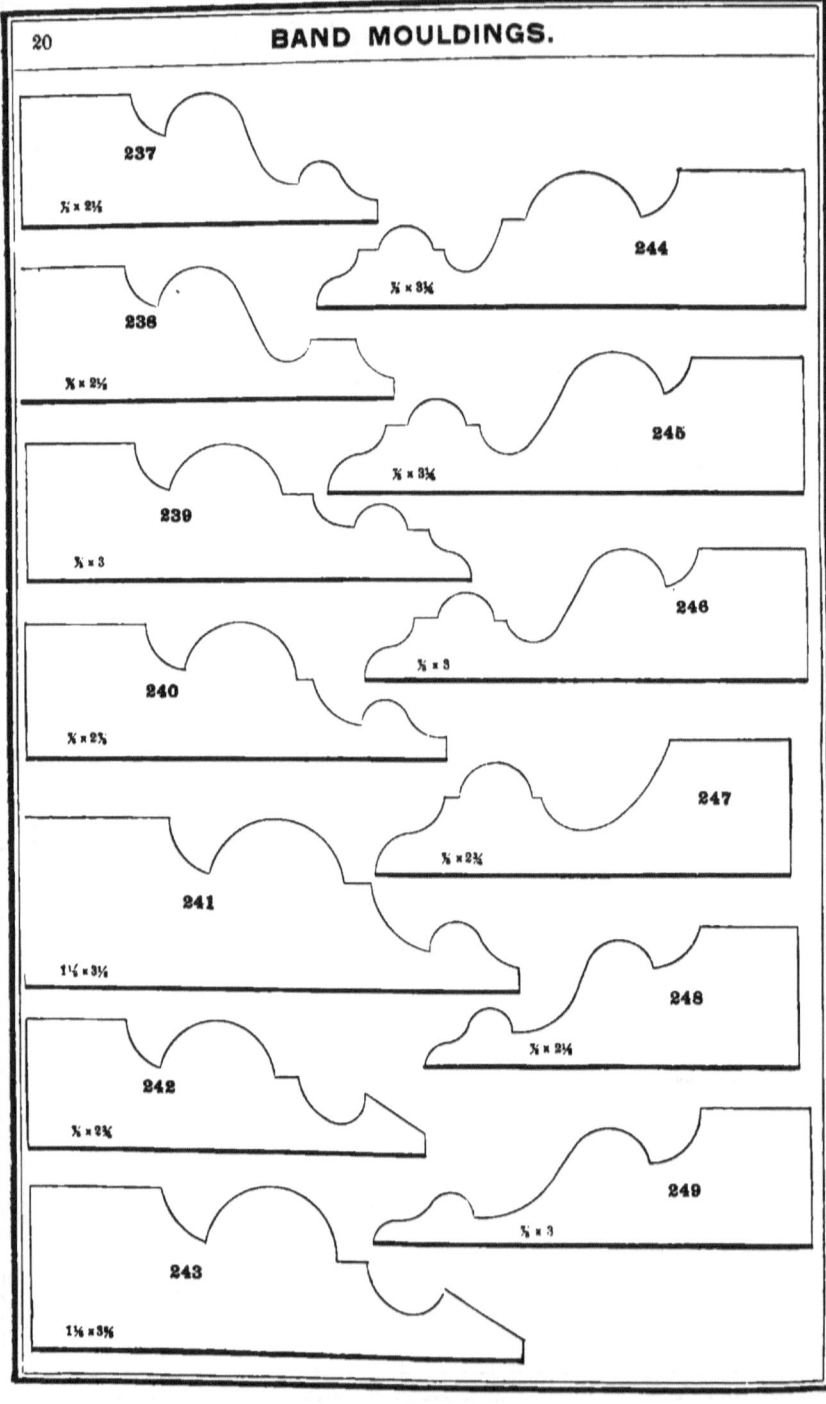

BAND MOULDINGS.

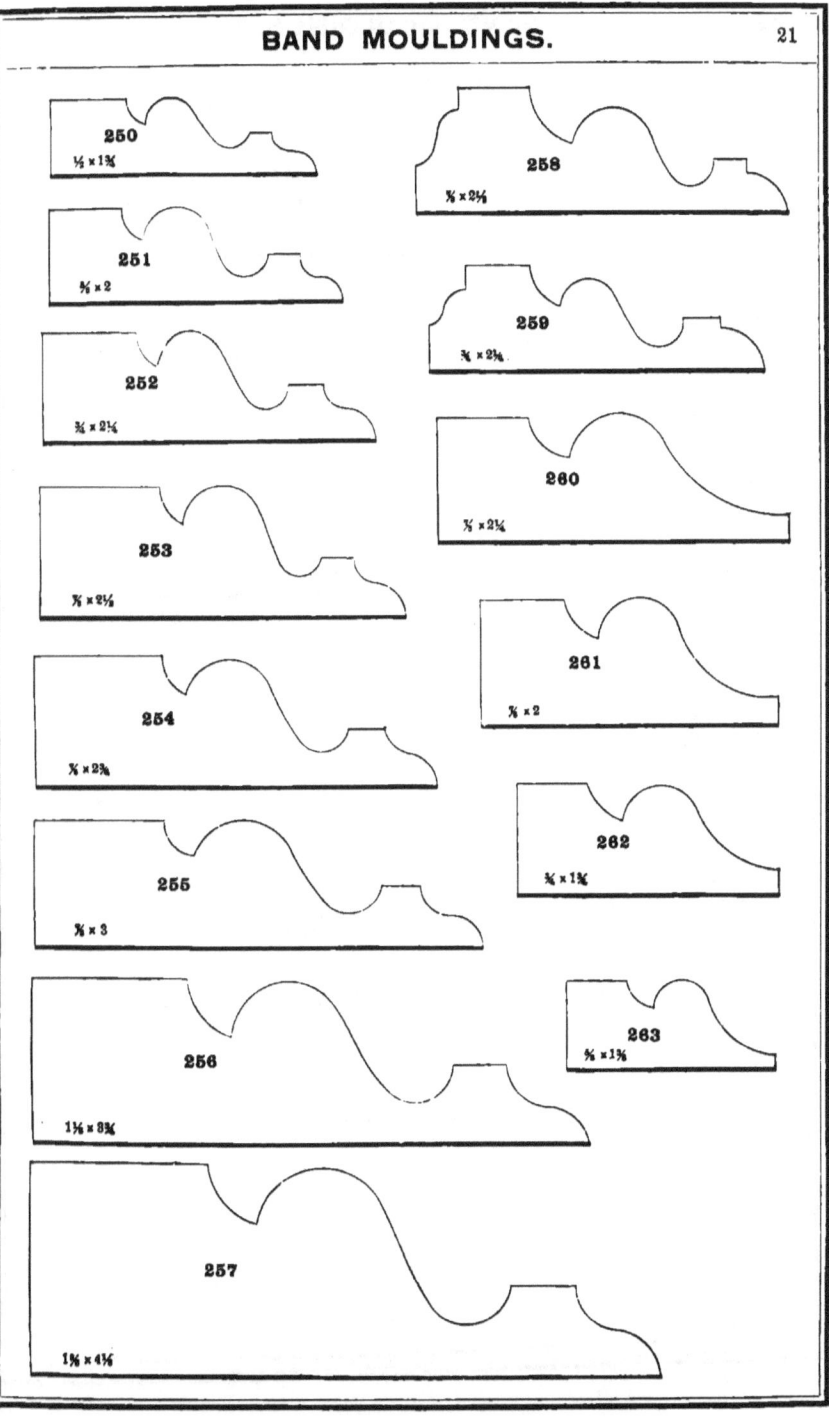

BAND MOULDINGS.

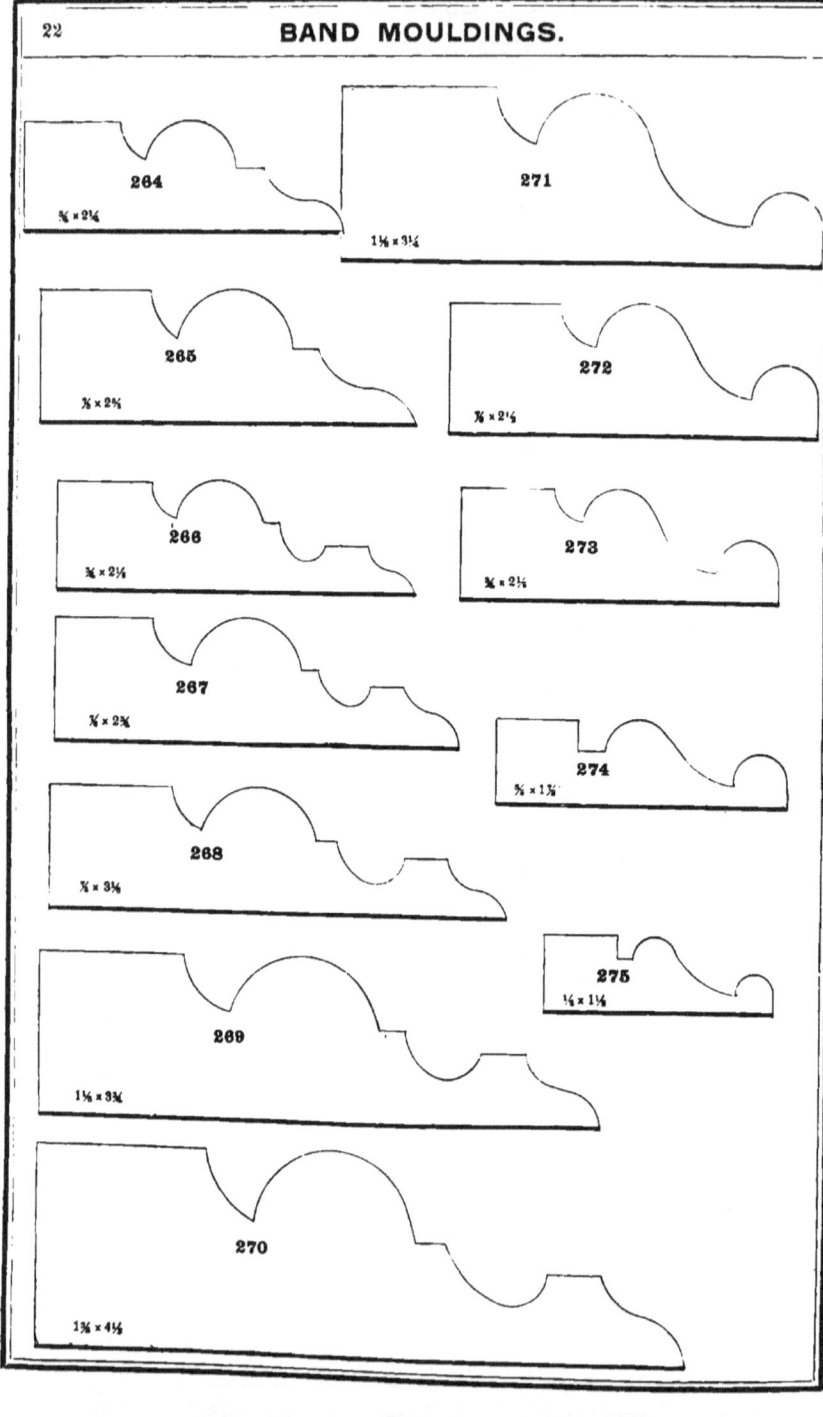

RABBETED PANEL AND BASE MOULDINGS.

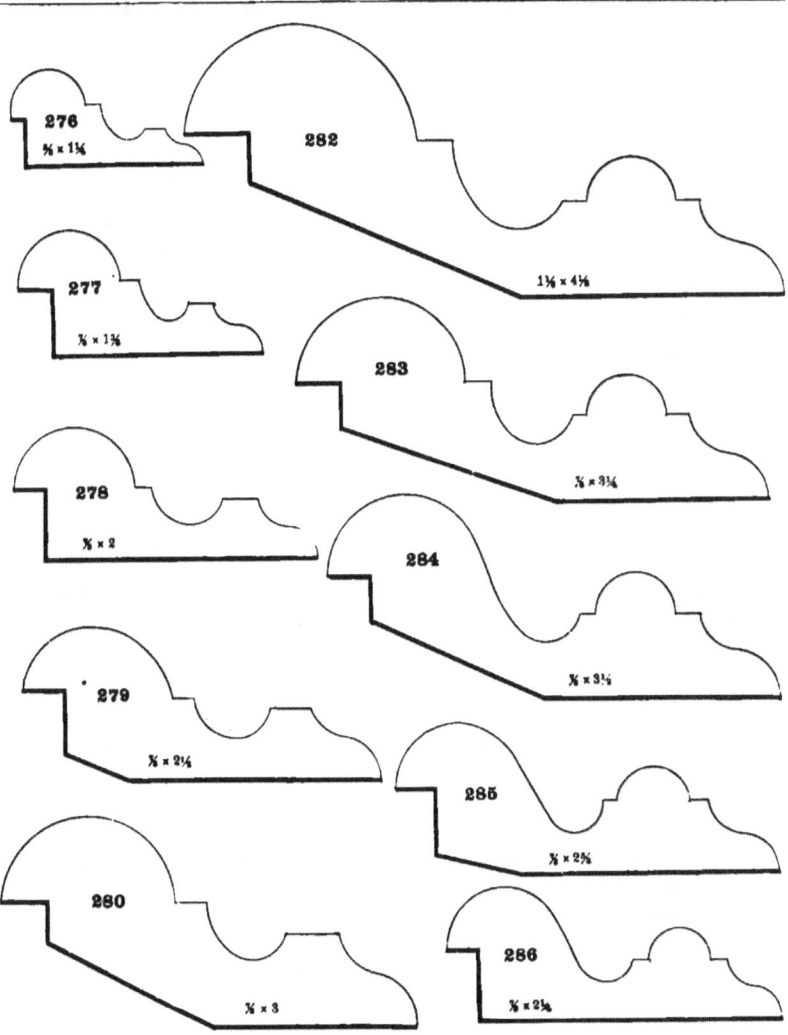

RABBETED PANEL AND BASE MOULDINGS.

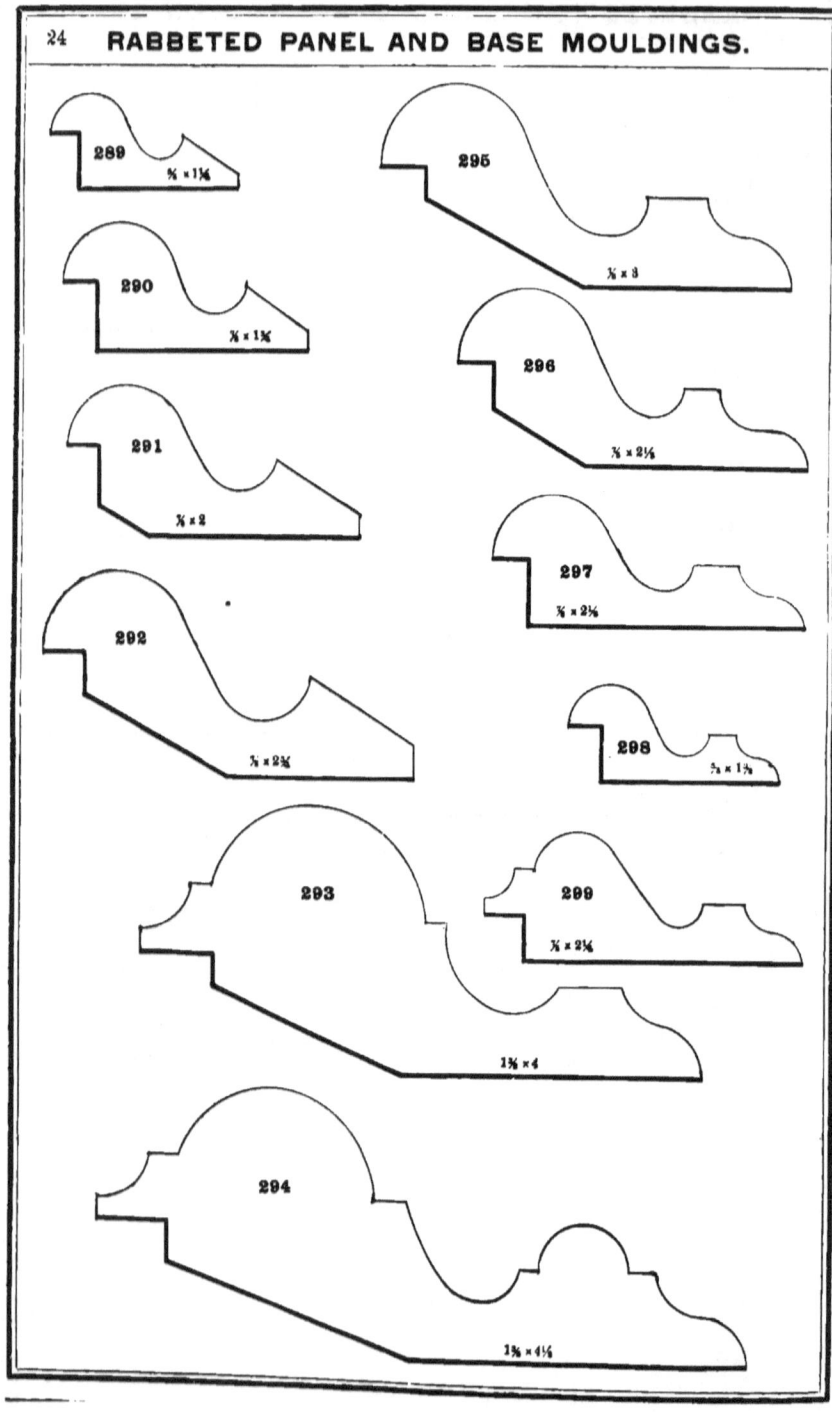

RABBETED PANEL AND BASE MOULDINGS.

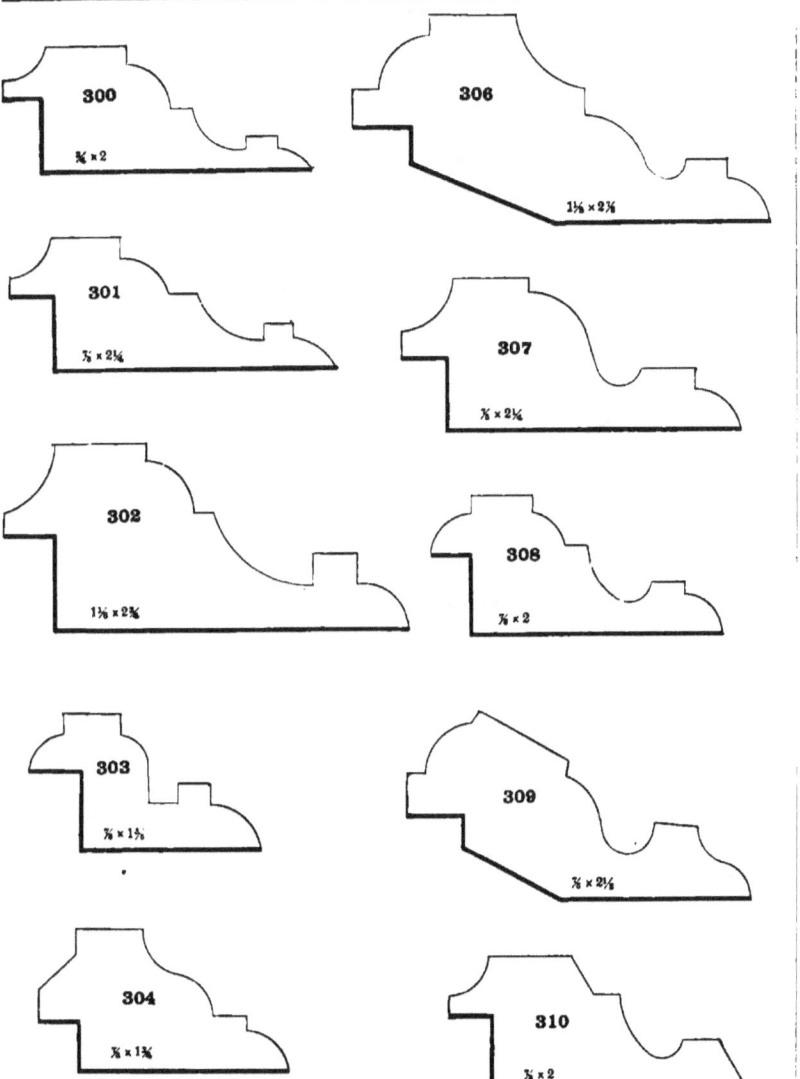

RABBETED PANEL AND BASE MOULDINGS.

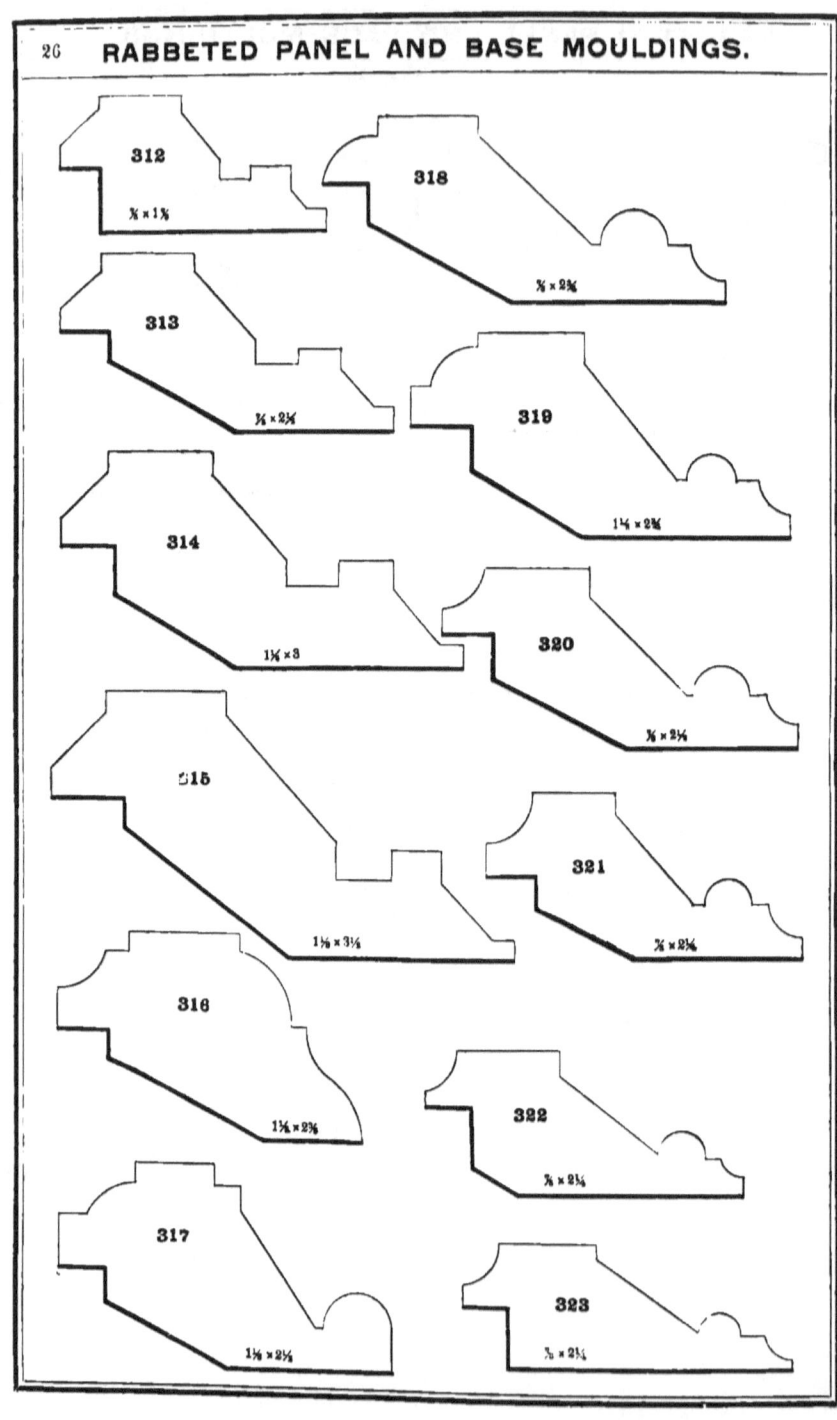

RABBETED PANEL AND BASE MOULDINGS.

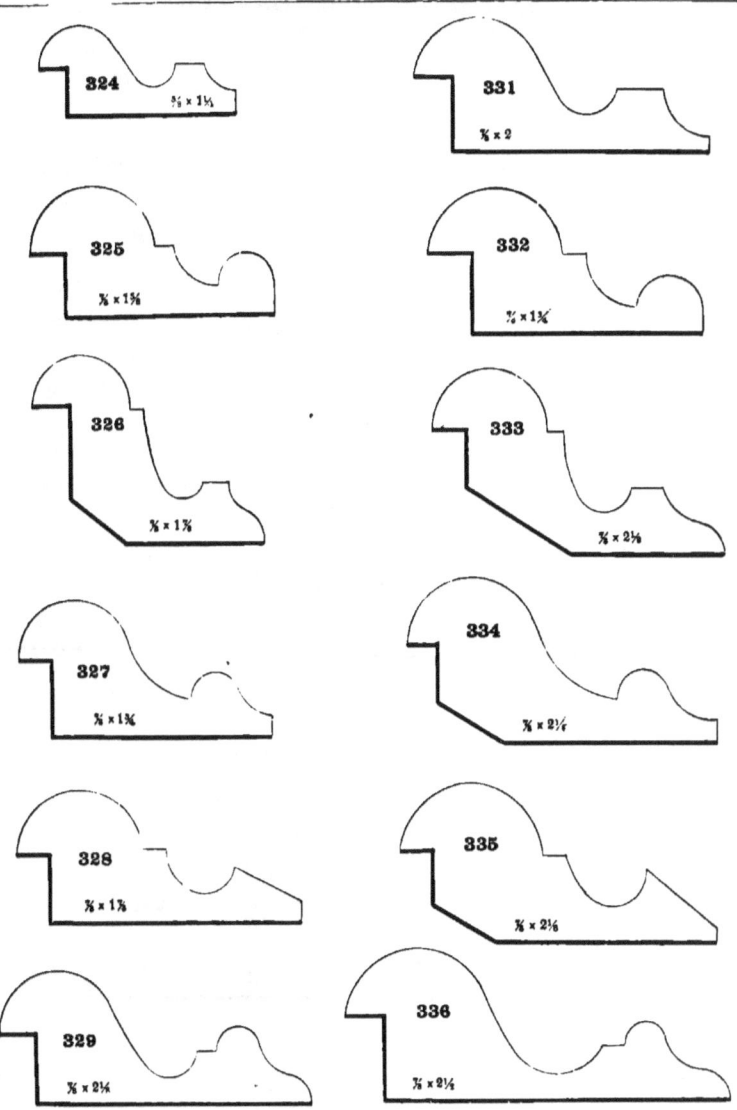

ASTRAGAL AND BATTENS.

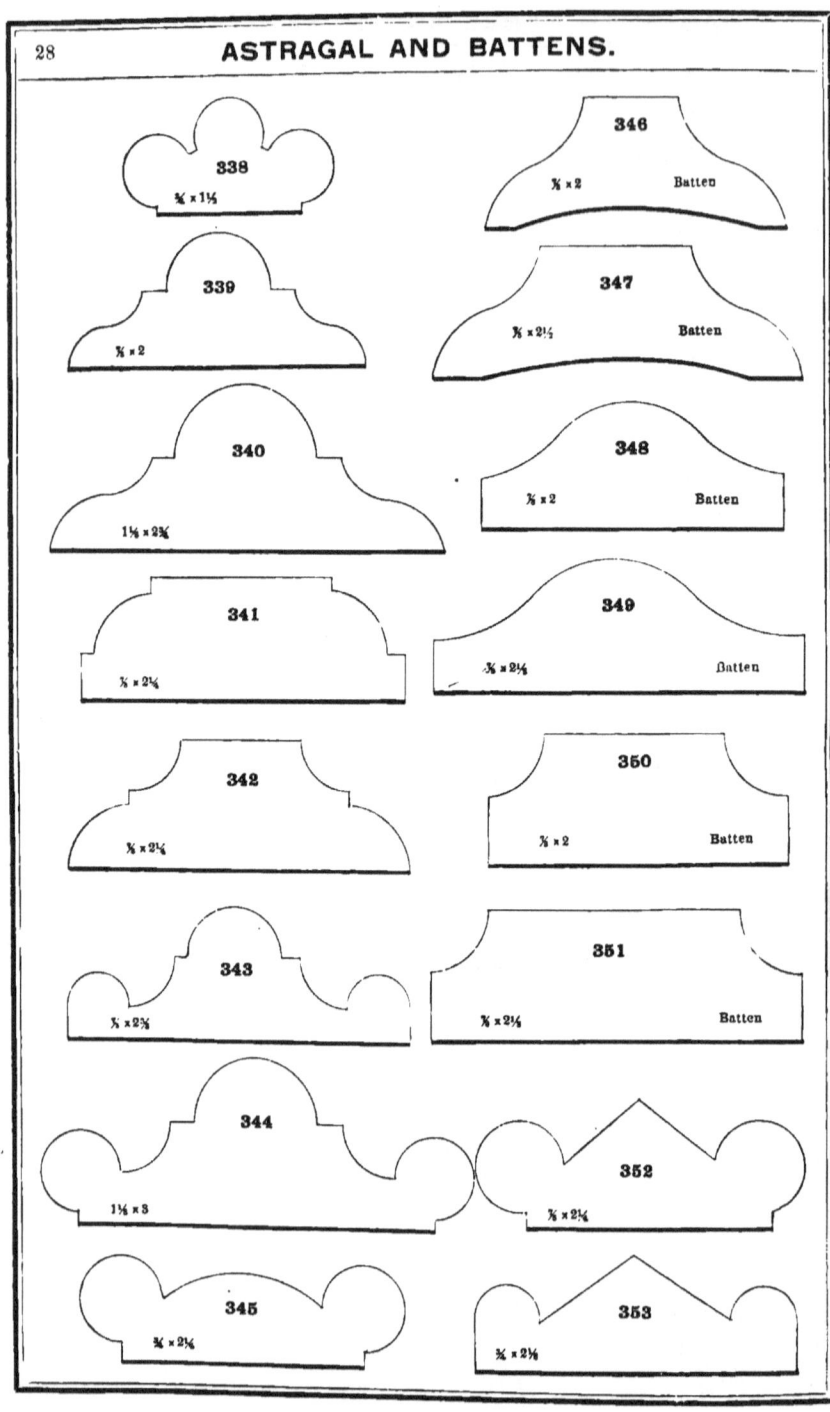

NOSINGS.

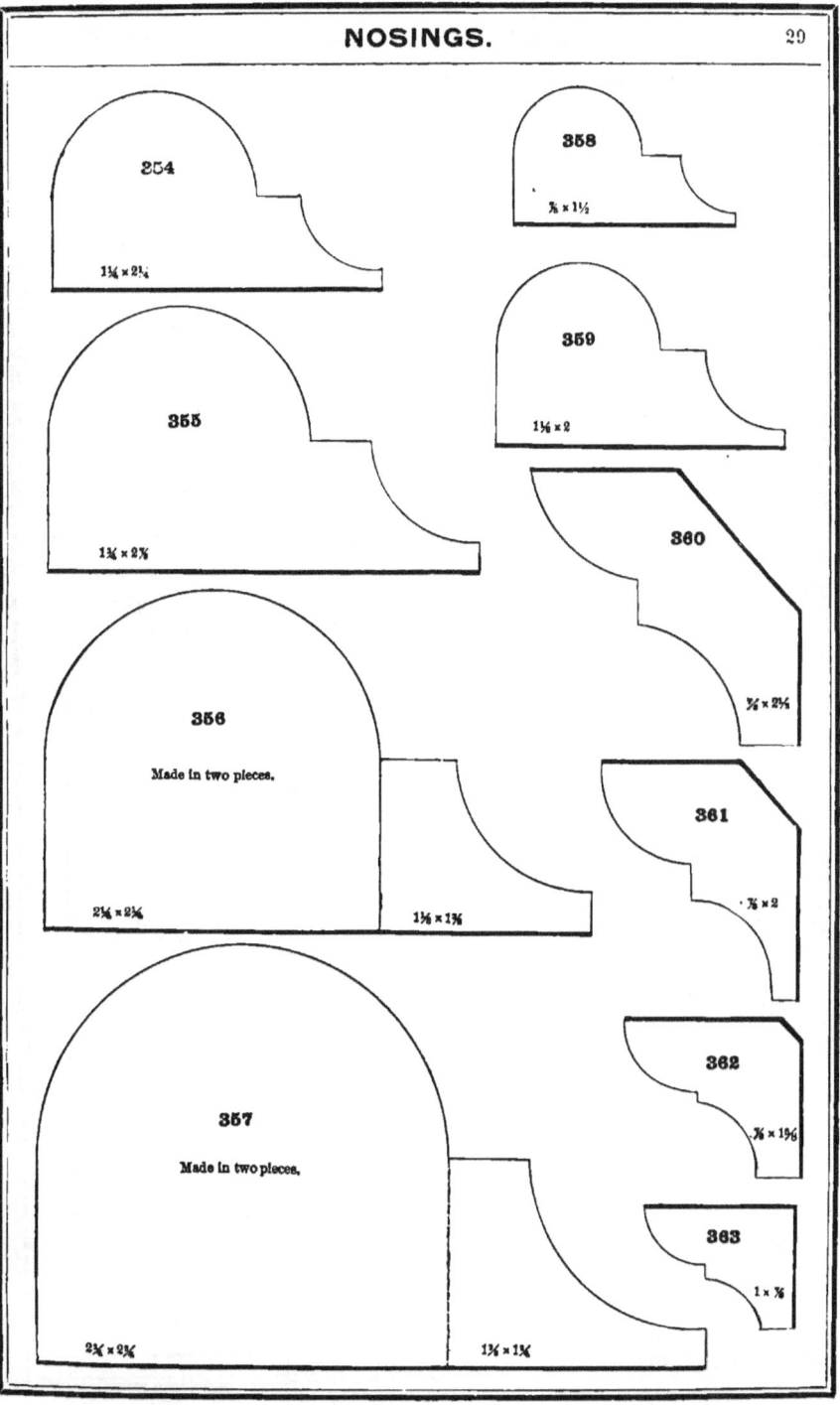

PEW BACK RAIL, WAINSCOTING CAP AND THRESHOLDS.

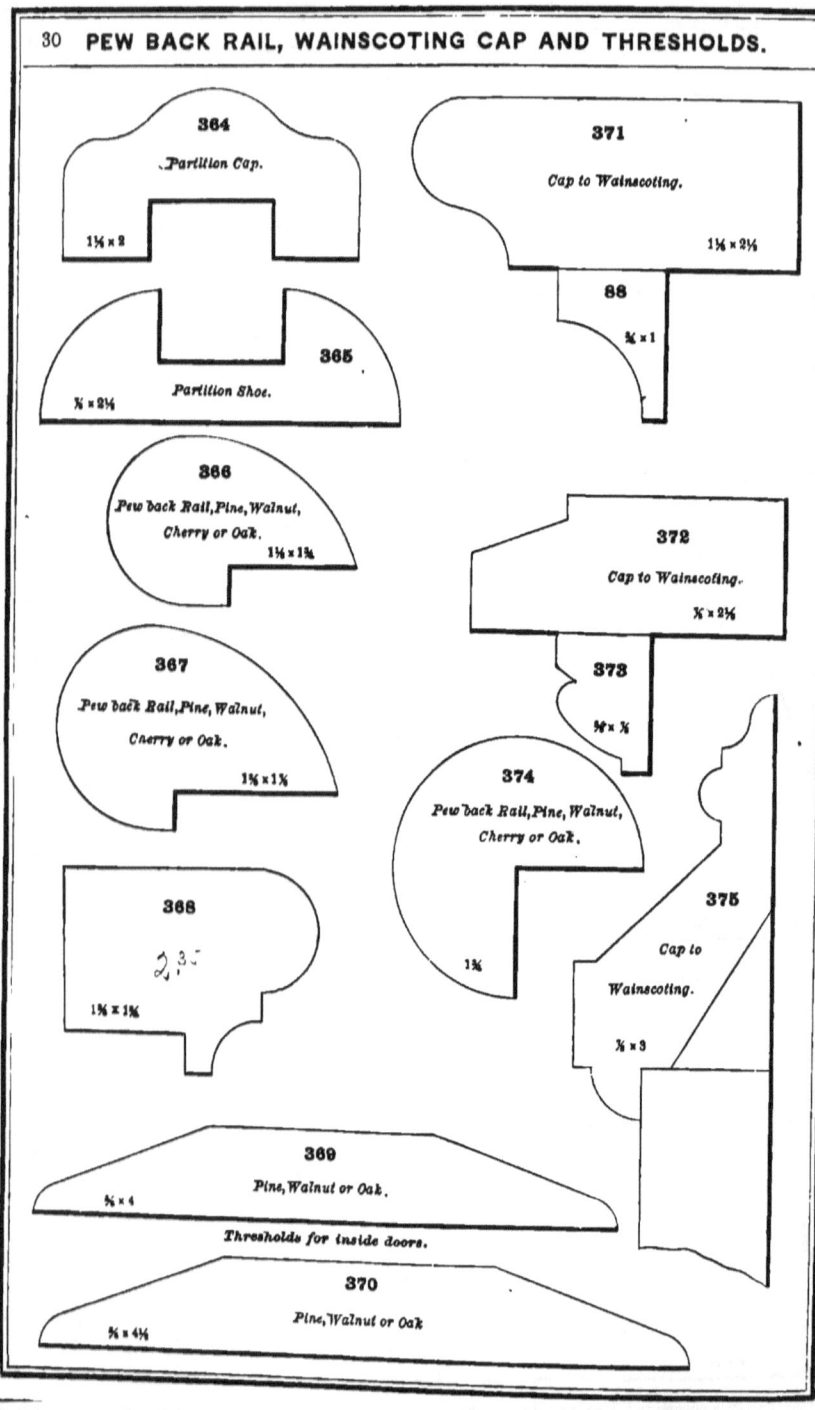

SUNK PANEL MOULDINGS.

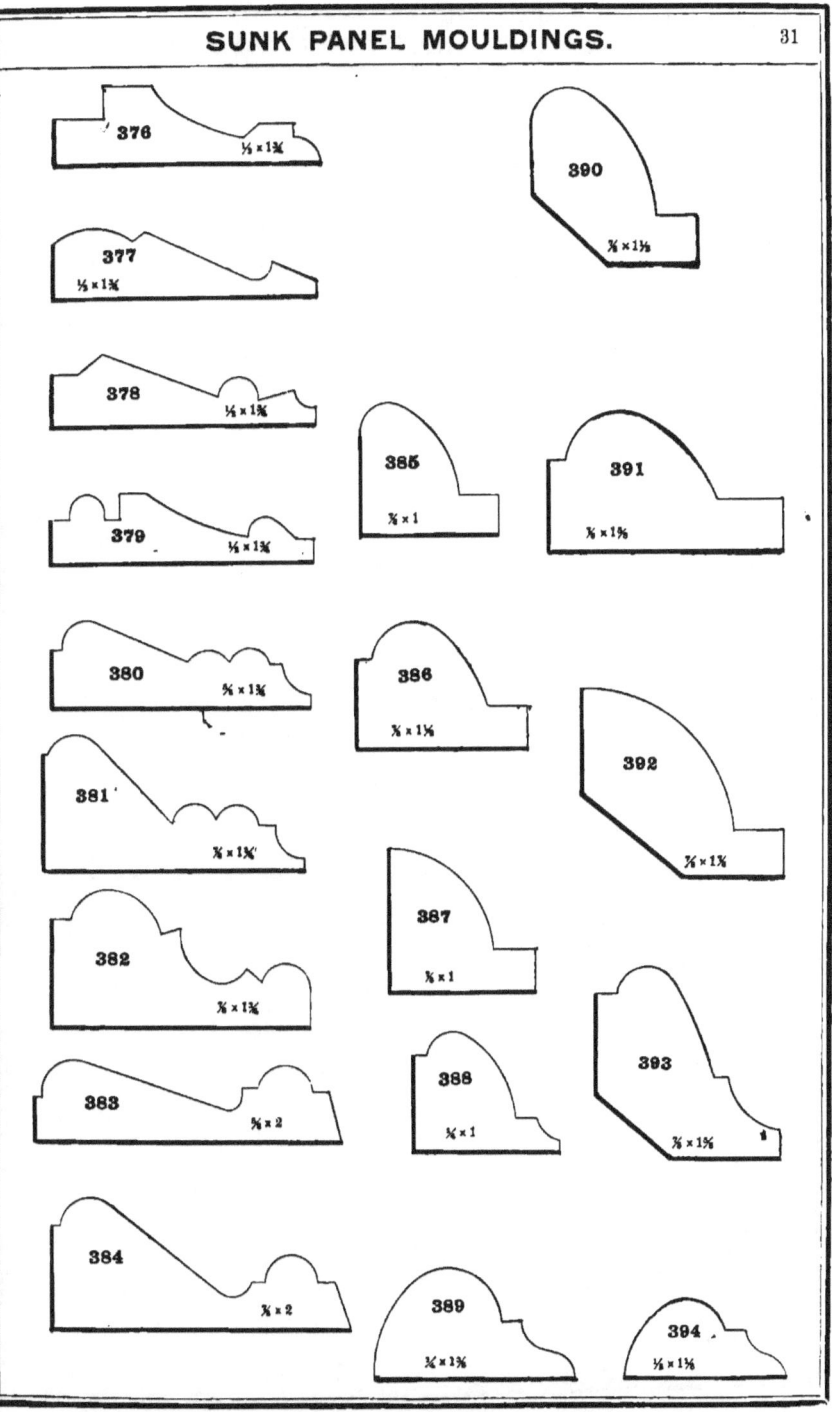

LATTICE, BACK BAND, AND TRANSOM BAR MOULDING.

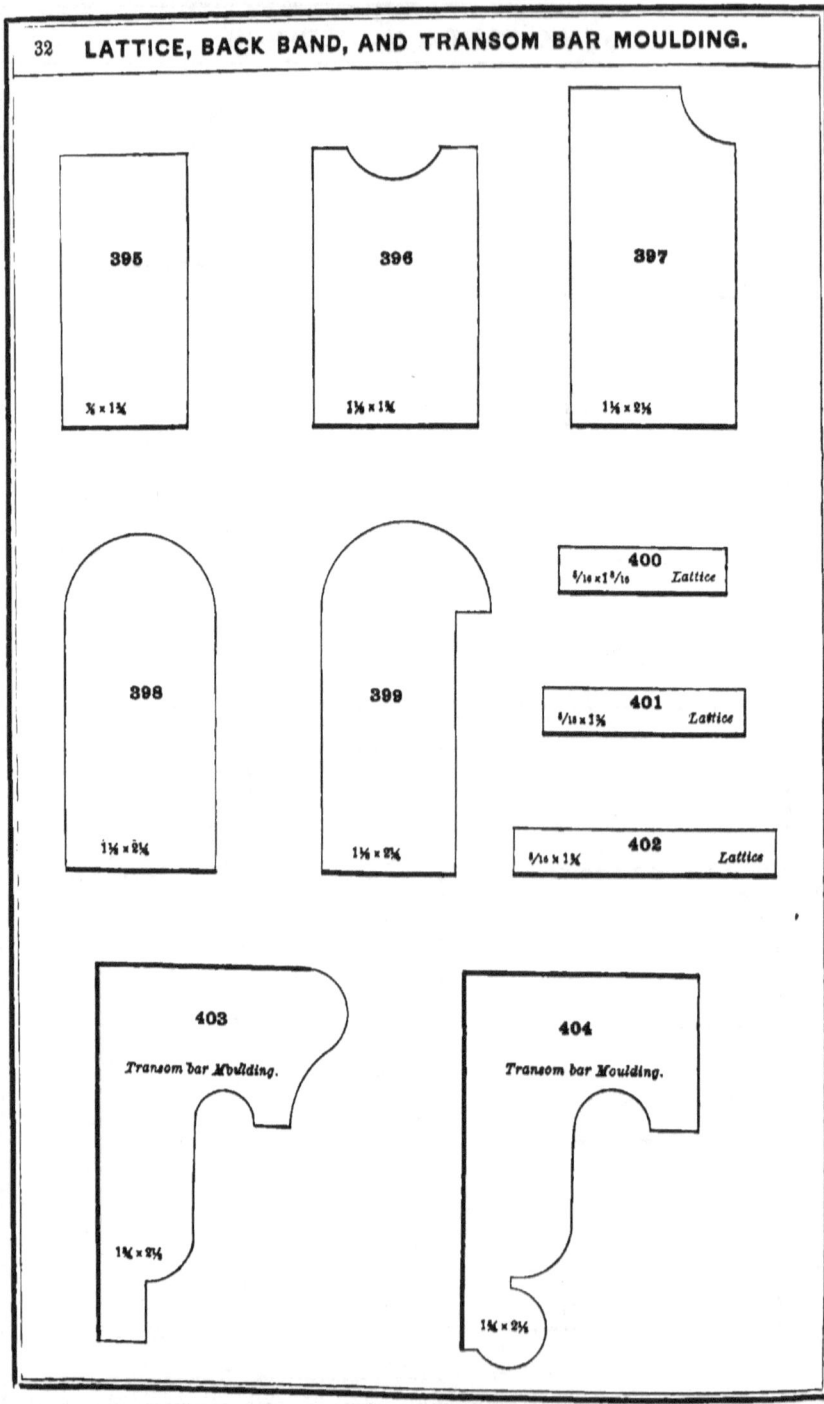

INTERIOR CORNICE.

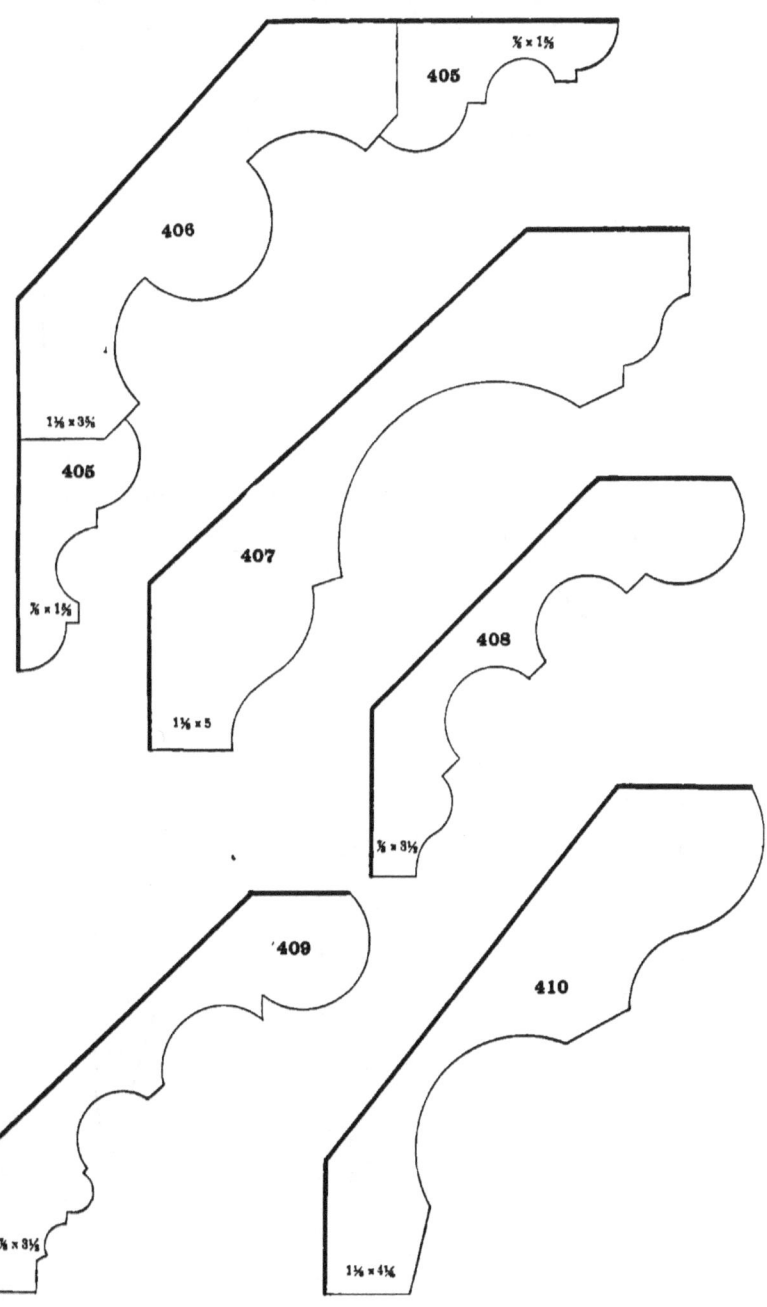

INTERIOR CORNICE AND BEAD MOULDINGS.

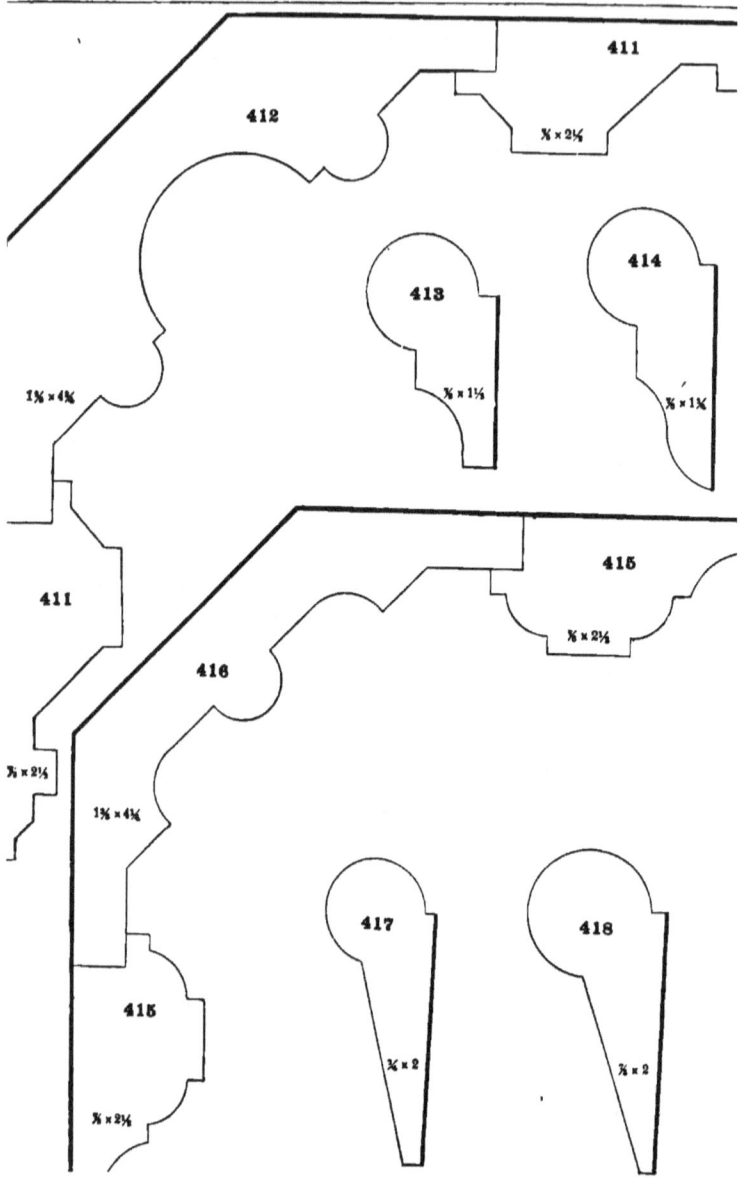

SECTION OF WINDOW FRAME.

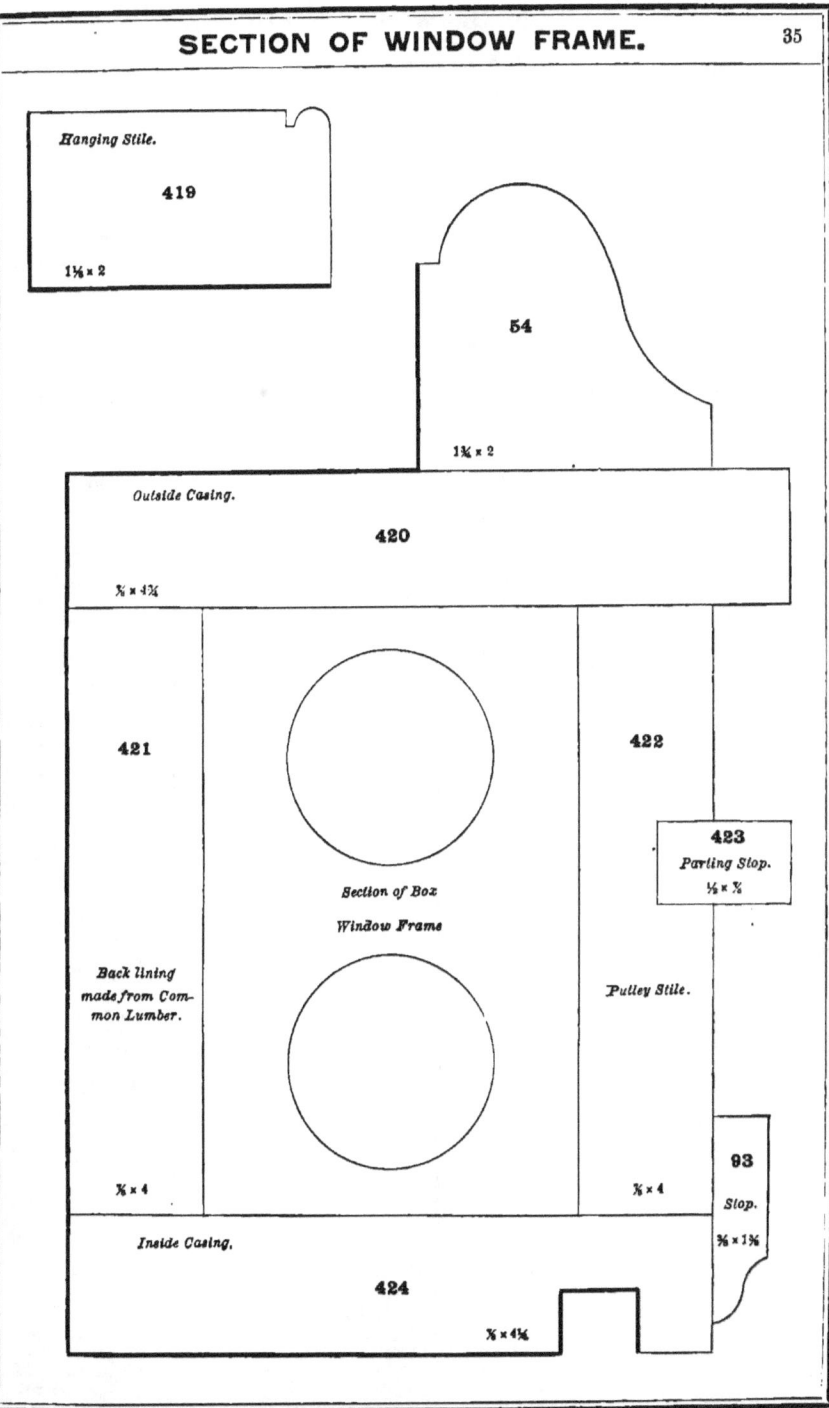

SECTION OF WINDOW FRAME.

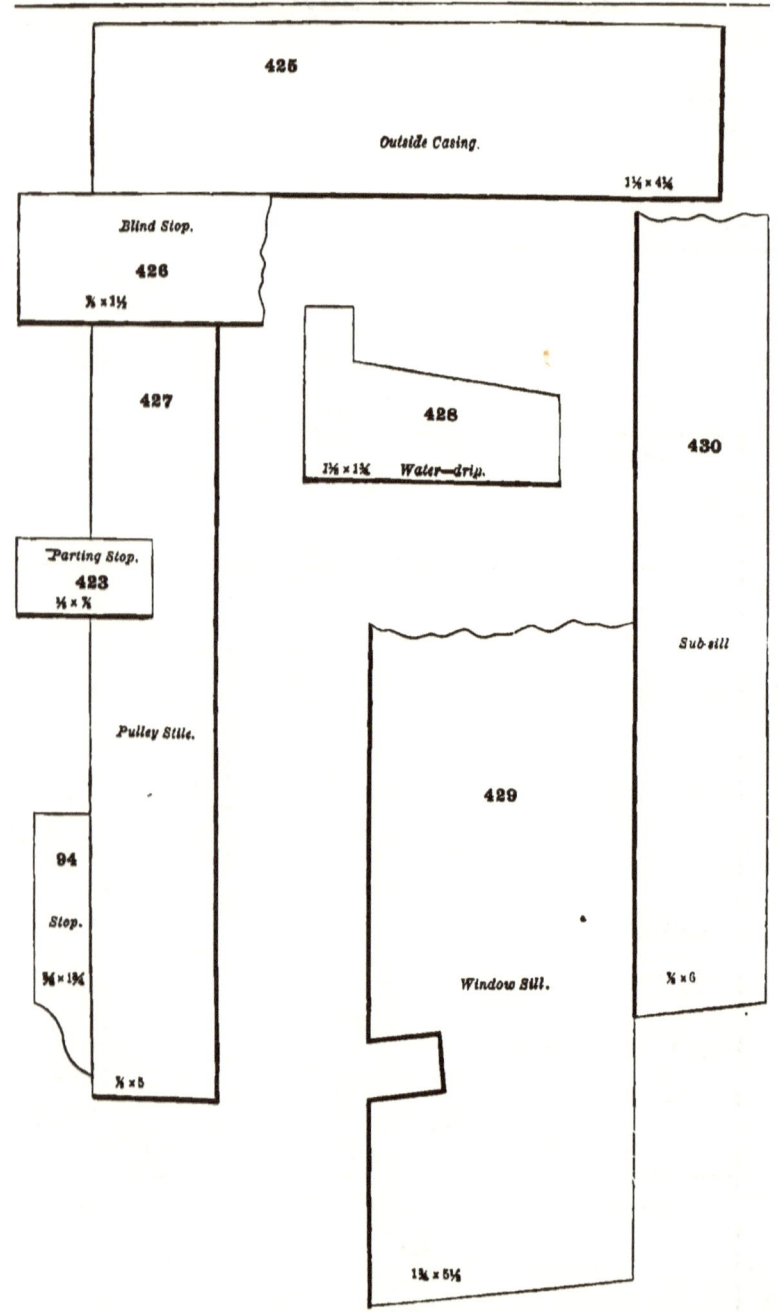

WATER TABLE OR DRIP CAP.

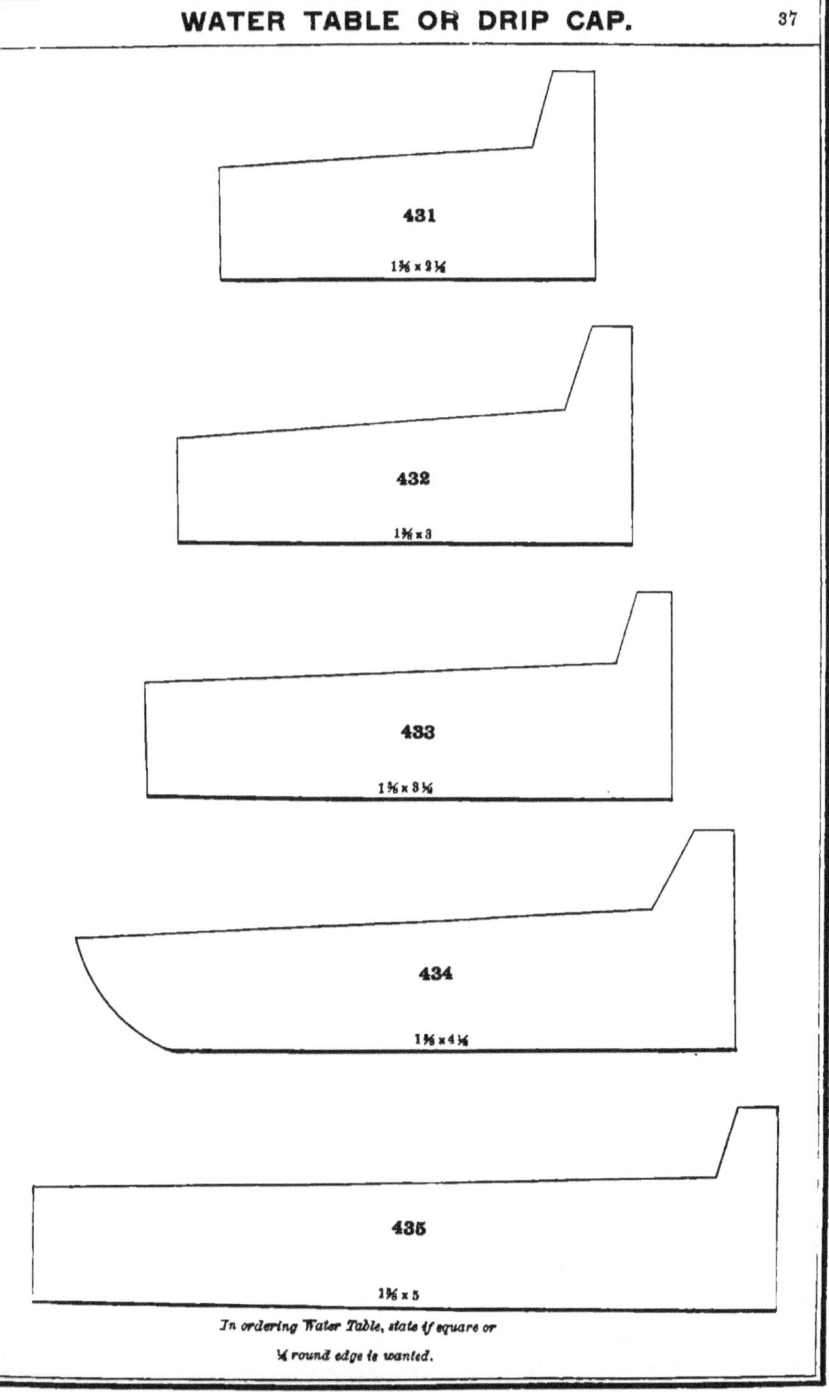

431
1⅝ x 2¼

432
1⅝ x 3

433
1⅝ x 3½

434
1⅝ x 4½

435
1⅝ x 5

In ordering Water Table, state if square or ¼ round edge is wanted.

CEILING AND WINDOW STOOLS.

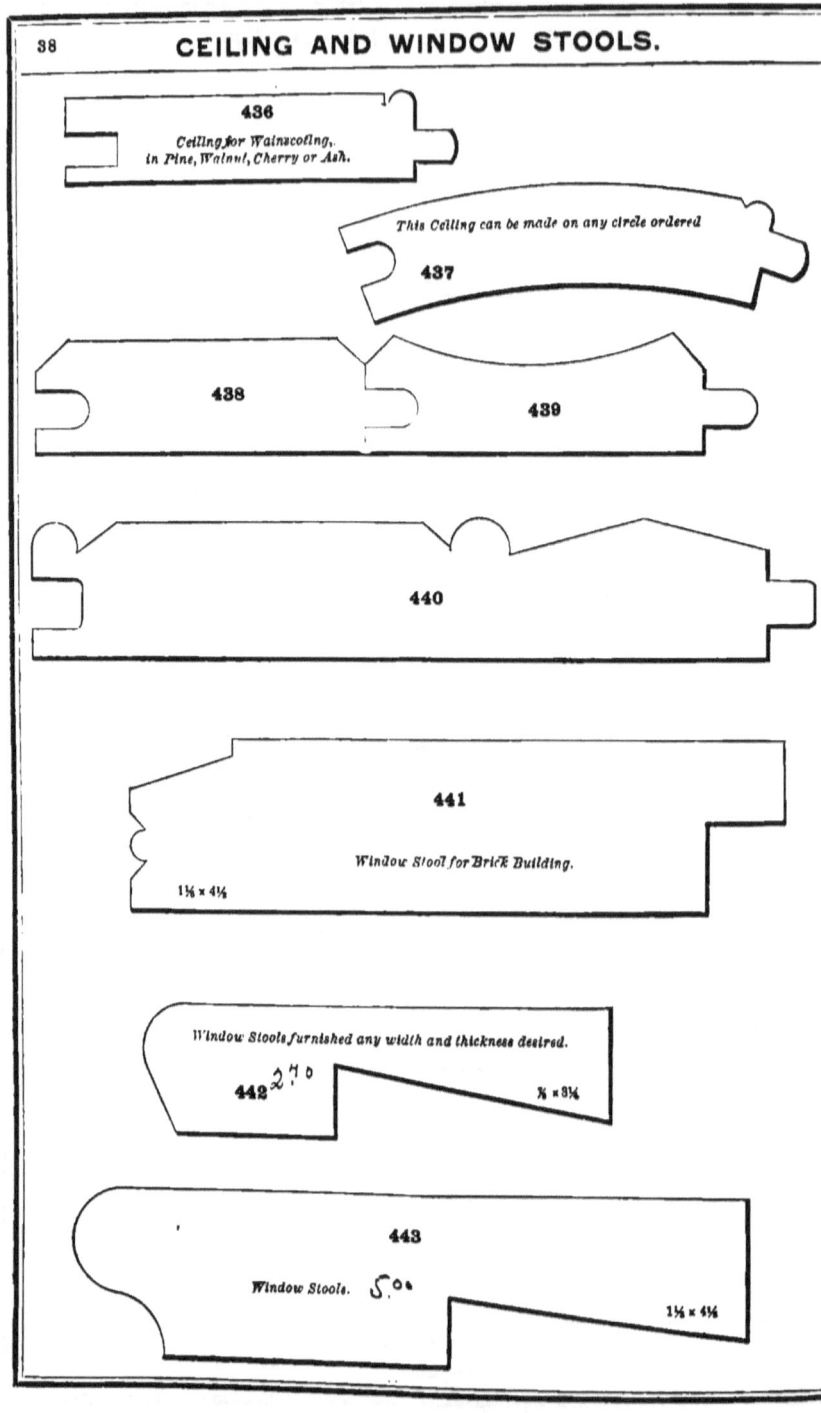

INSIDE FINISH.

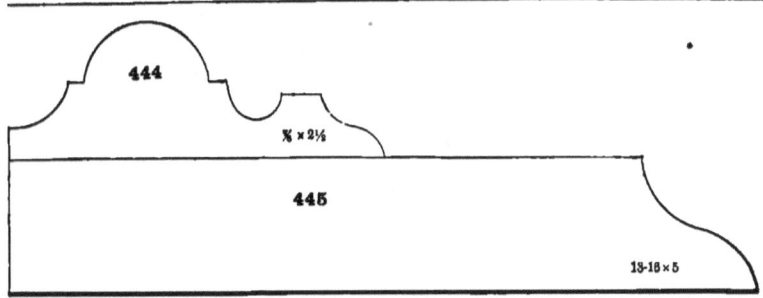

444
⅞ x 2½
445
13-16 x 5

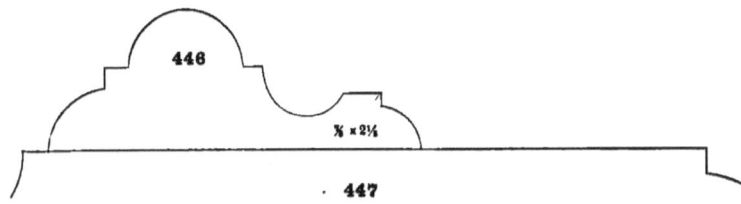

446
⅞ x 2½
447

INSIDE FINISH.

⅞ × 2¾

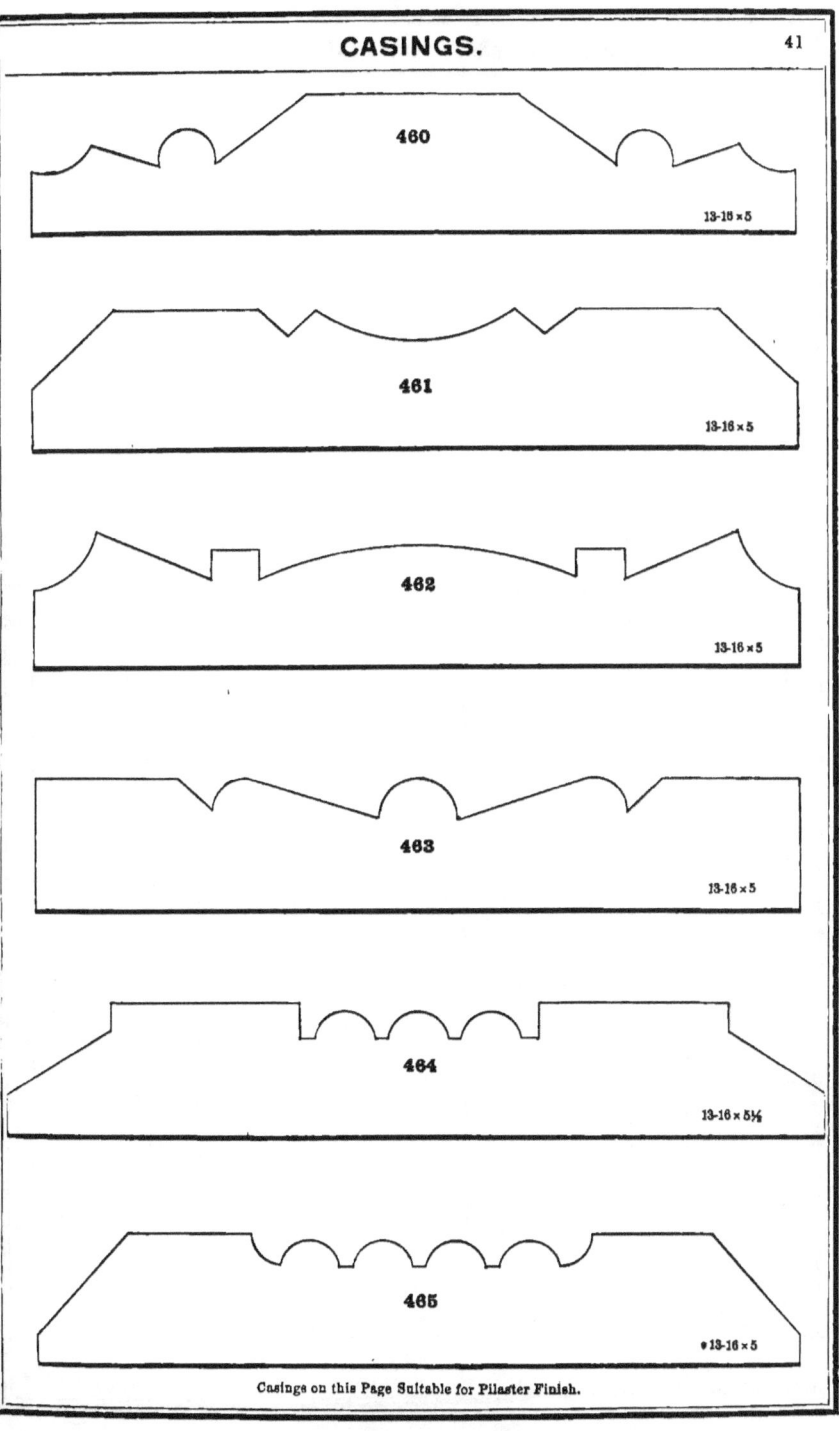

EASTLAKE AND QUEEN ANNE CASINGS FOR WINDOWS AND DOORS.

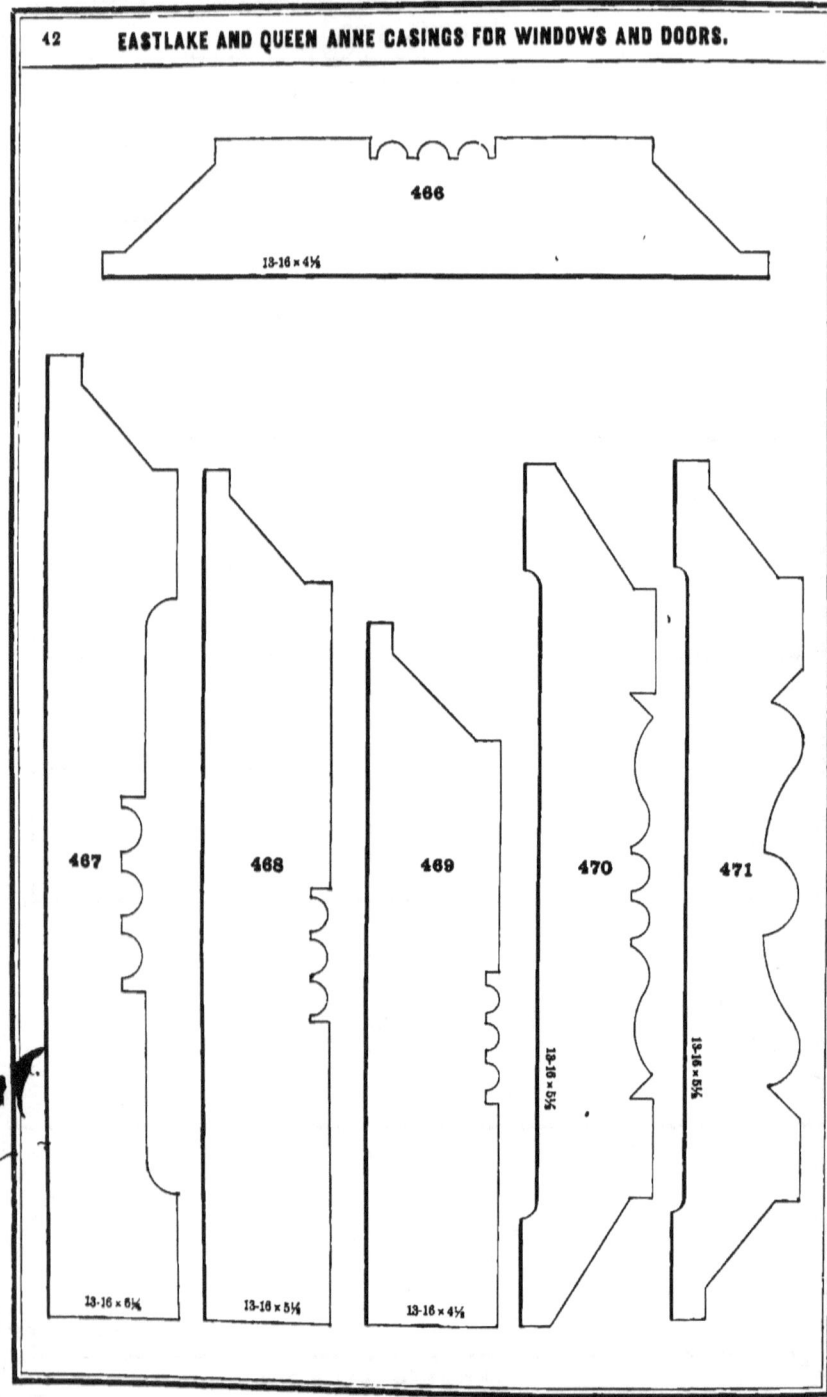

EASTLAKE AND QUEEN ANNE CASINGS FOR WINDOWS AND DOORS.

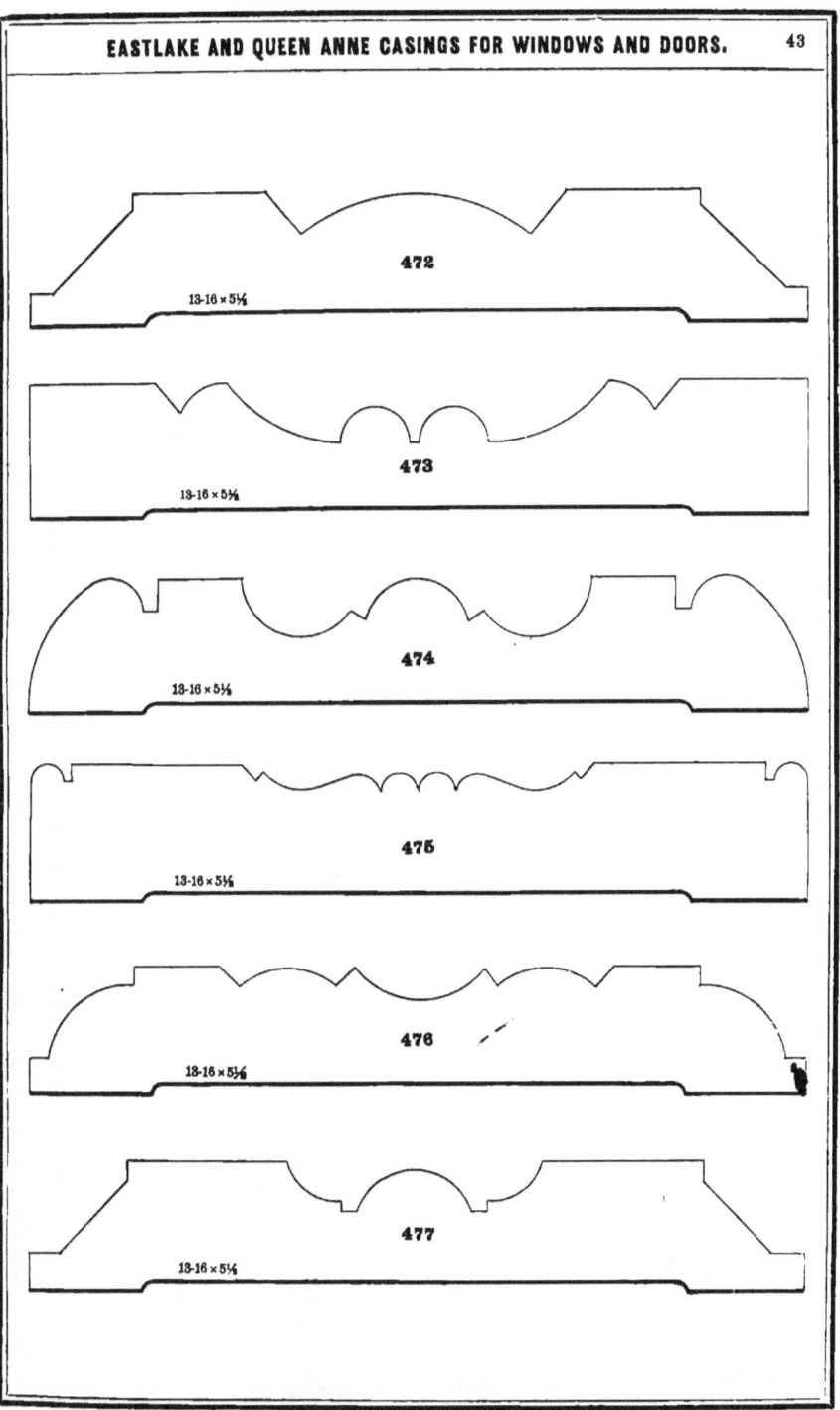

CASINGS.

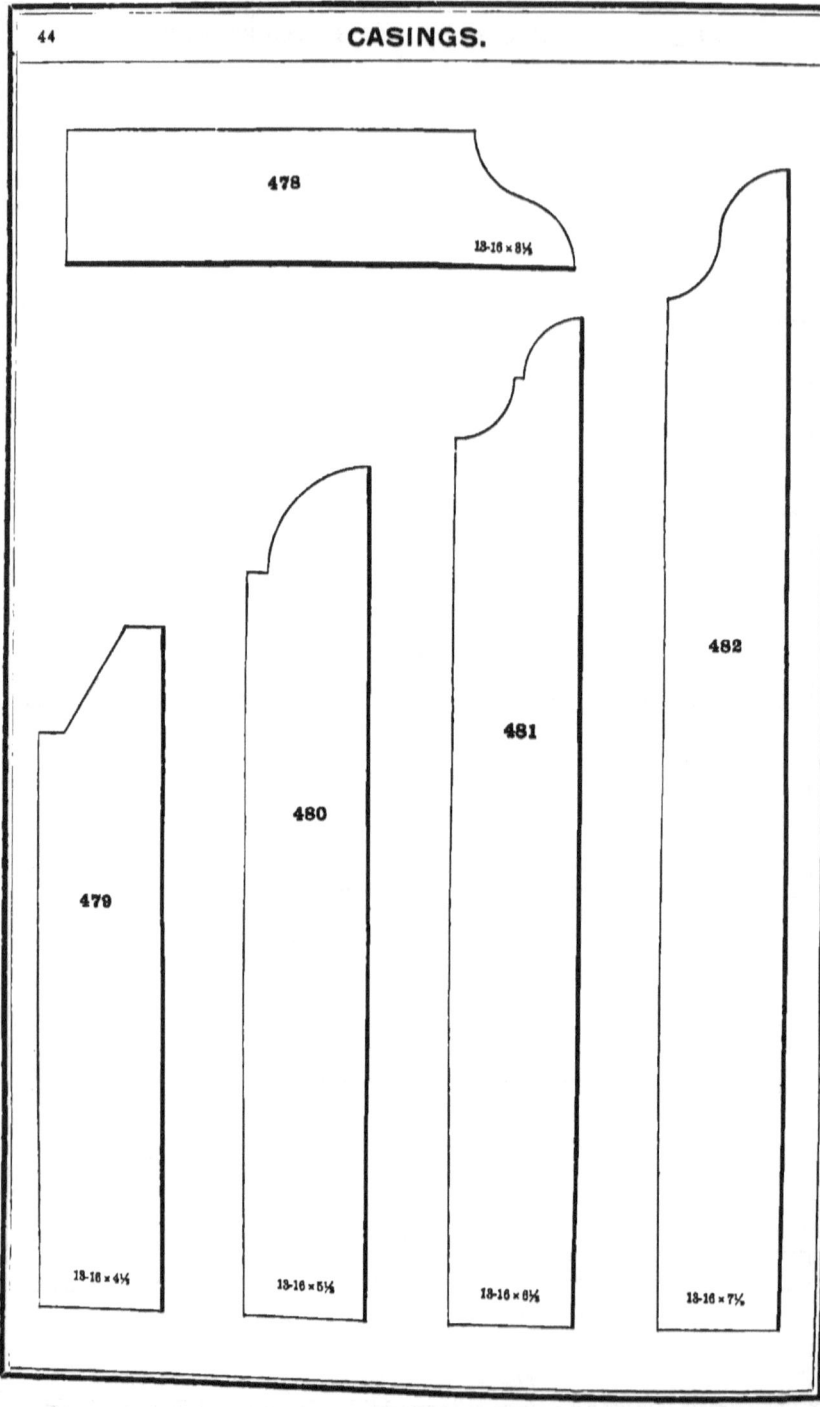

BASE.

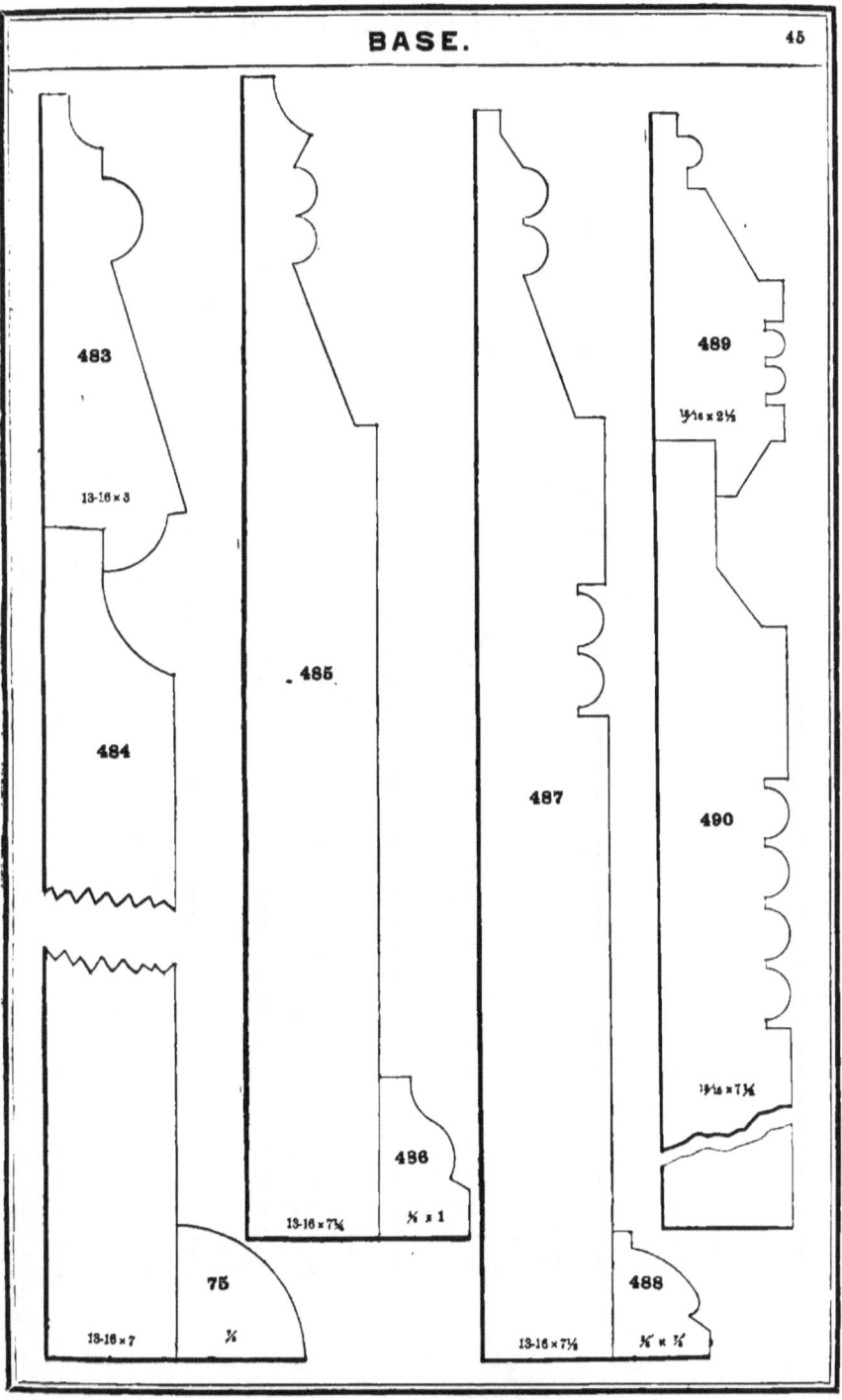

DROP SIDING, FLOORING AND SHIP LAP.

Flooring; in Walnut, Cherry or Ash.

491

DROP SIDING.

492 **493** **494** **495**

CORNICE DRAPERY.

BRACKETS.

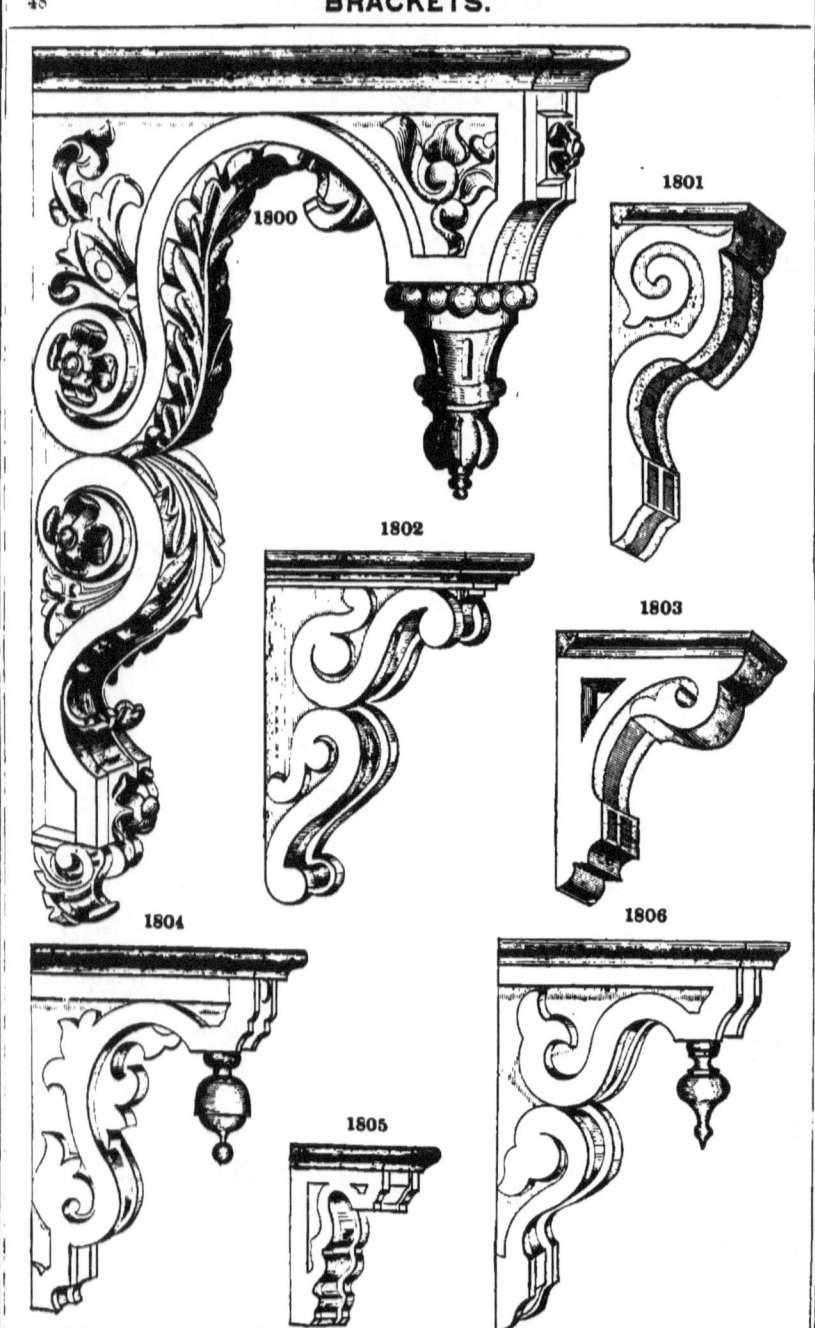

BRACKETS.

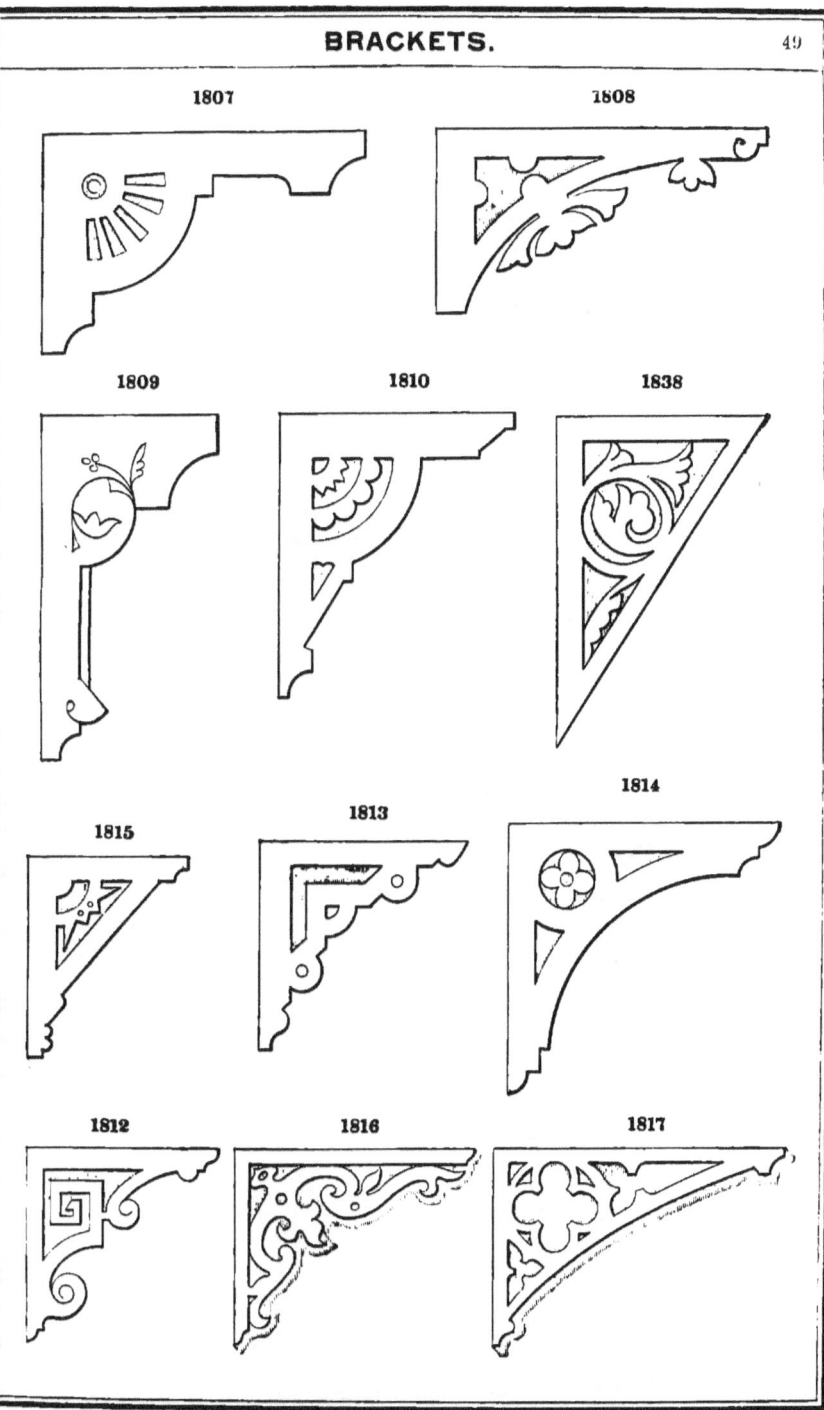

BRACKETS.

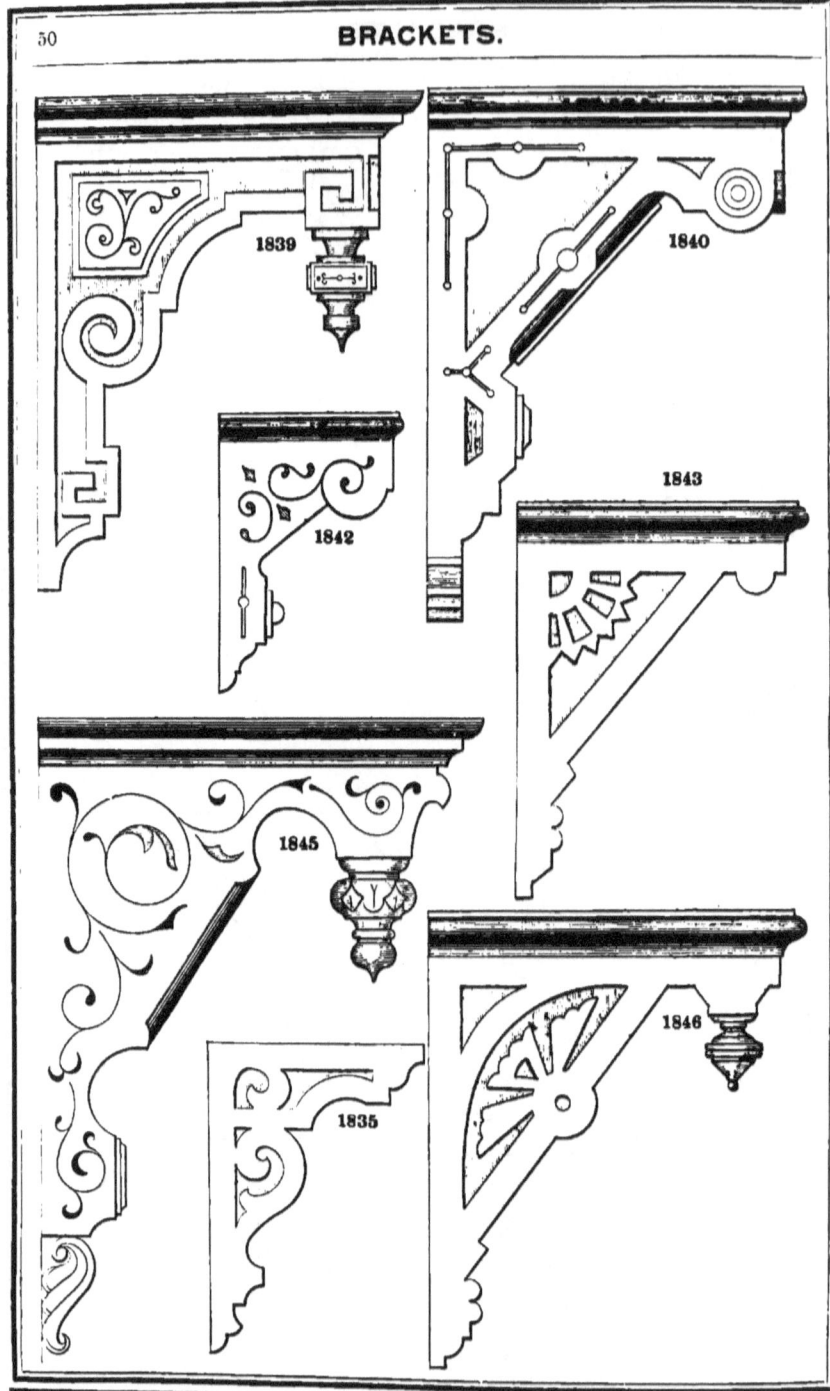

BRACKETS.

BRACKETS.

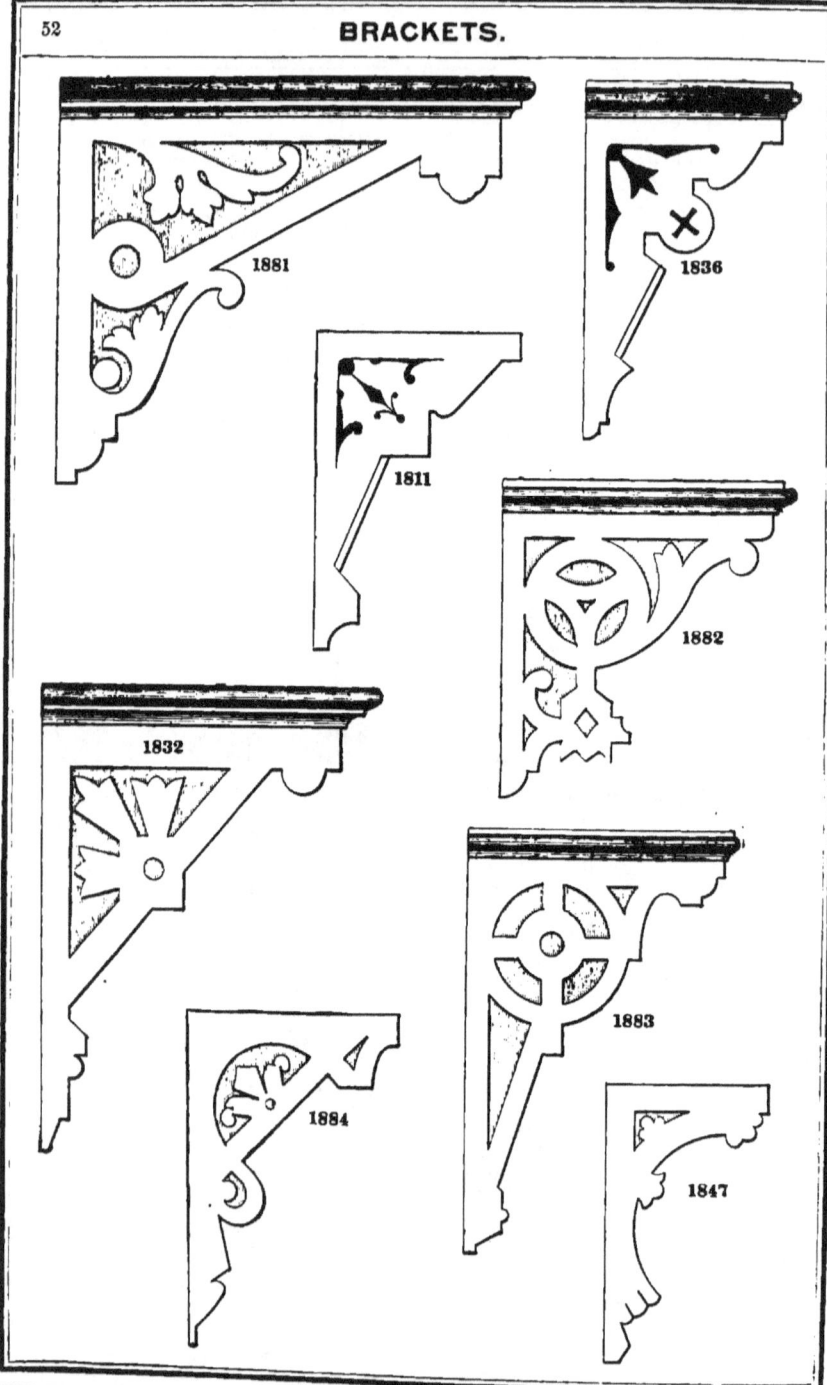

BRACKETS.

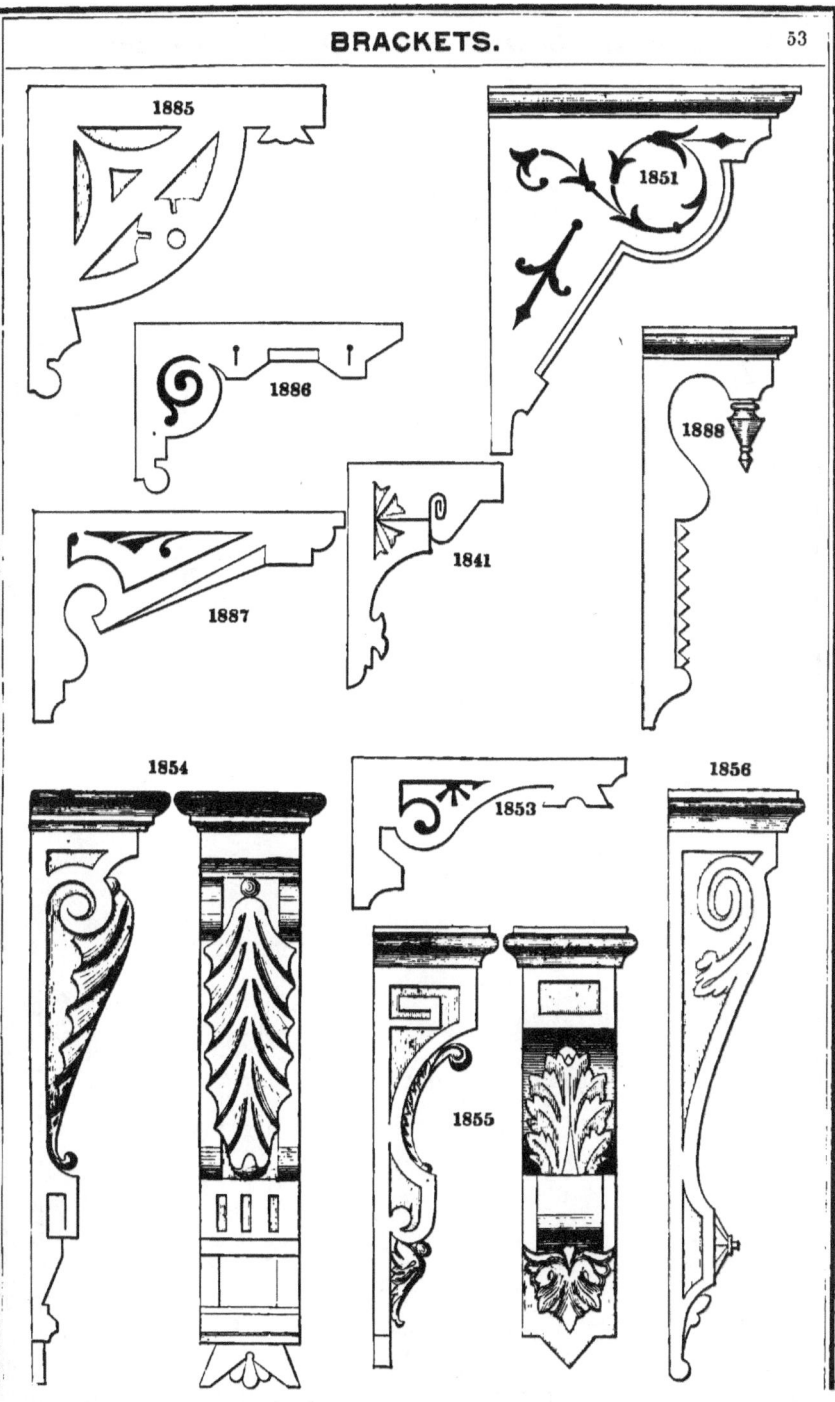

VERANDA SAWED BALUSTRADE AND RAIL.

1710 1711 1712

1604
Pine Rail,
for Sawed Balusters
$7.50 per hundred feet.

1610
Foot Rail for Sawed Balustrade $3.00 per hundred feet.

1713 1714 1715

OUTSIDE RAIL, BALUSTERS, AND POSTS. 55

1601

Pine Rail for Outside.
5½ in. wide $13.75 per hundred feet.
7 ,, ,, 17.50 ,, ,, ,,

$1.00 per hundred feet.

1639 1666 1667 1668 1663 1664 1665 1640

PICKETS, GATE, AND FENCE.

| 2365 | 2366 |  2367 |  2368 | 2369 |  2370 |

2375   2376

2377   2378

VERANDAS.

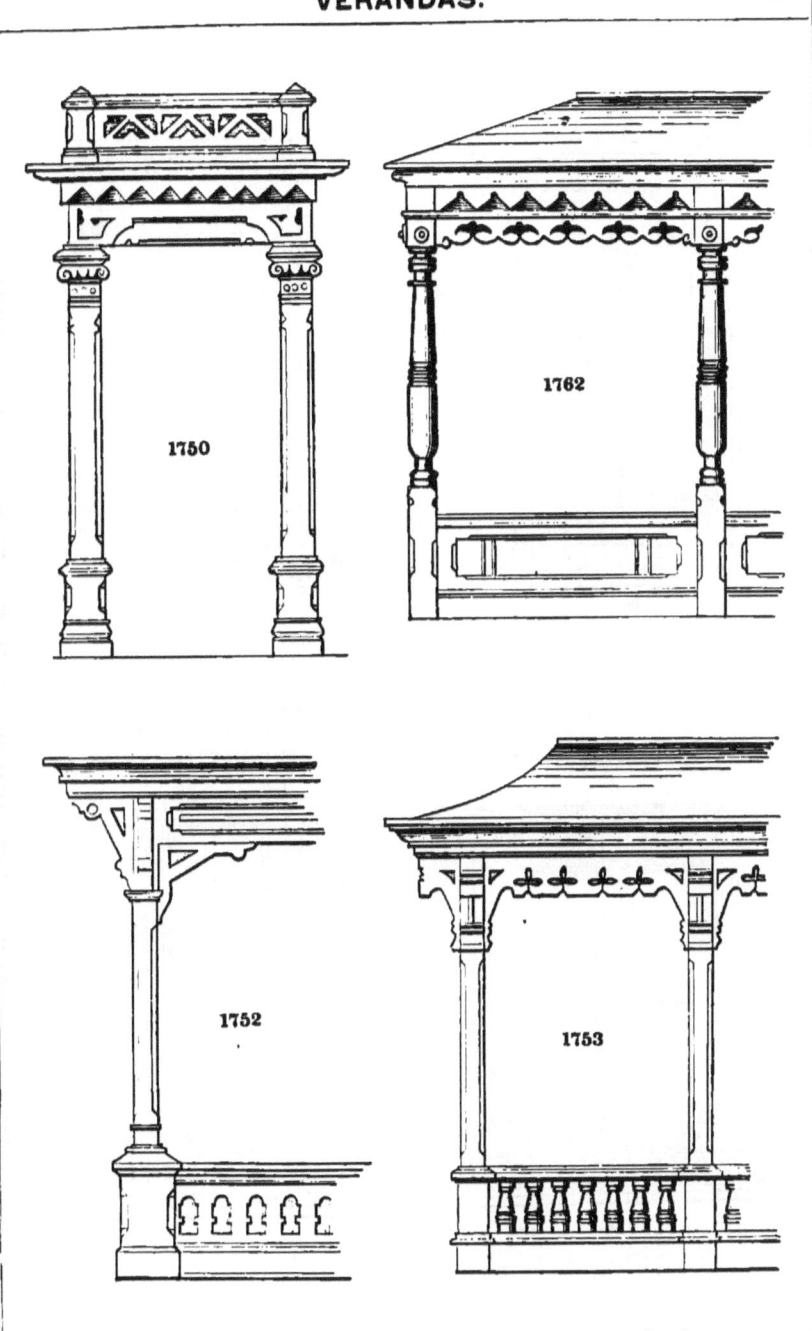

BAY WINDOWS.

1780

1781

1782

1783

PEW ENDS AND PULPITS.

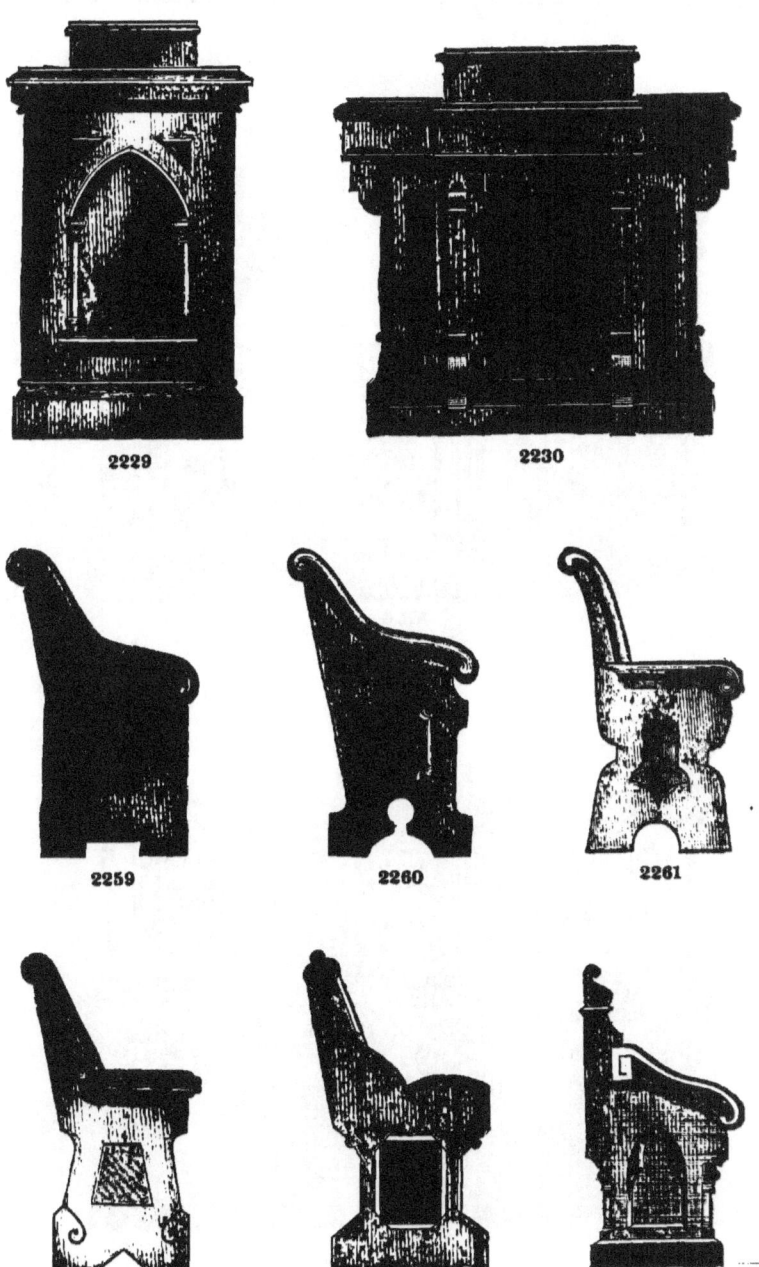

2229 2230

2259 2260 2261

2262 2263 2264

OFFICE OR BANK COUNTERS.

2200

2201

WOOD MANTELS.

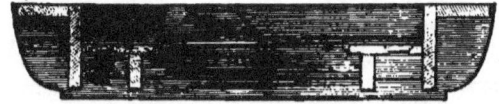

Mantels made of Pine, Oak, Cherry, Walnut, or Mahogany.

STORE FRONTS.

Left.   **452**   Right.

Left.   Right.

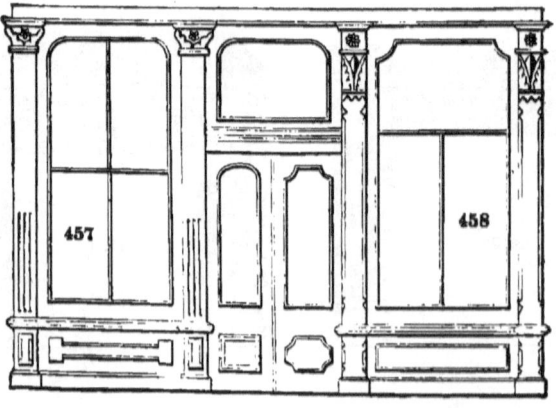

STORE DOORS.

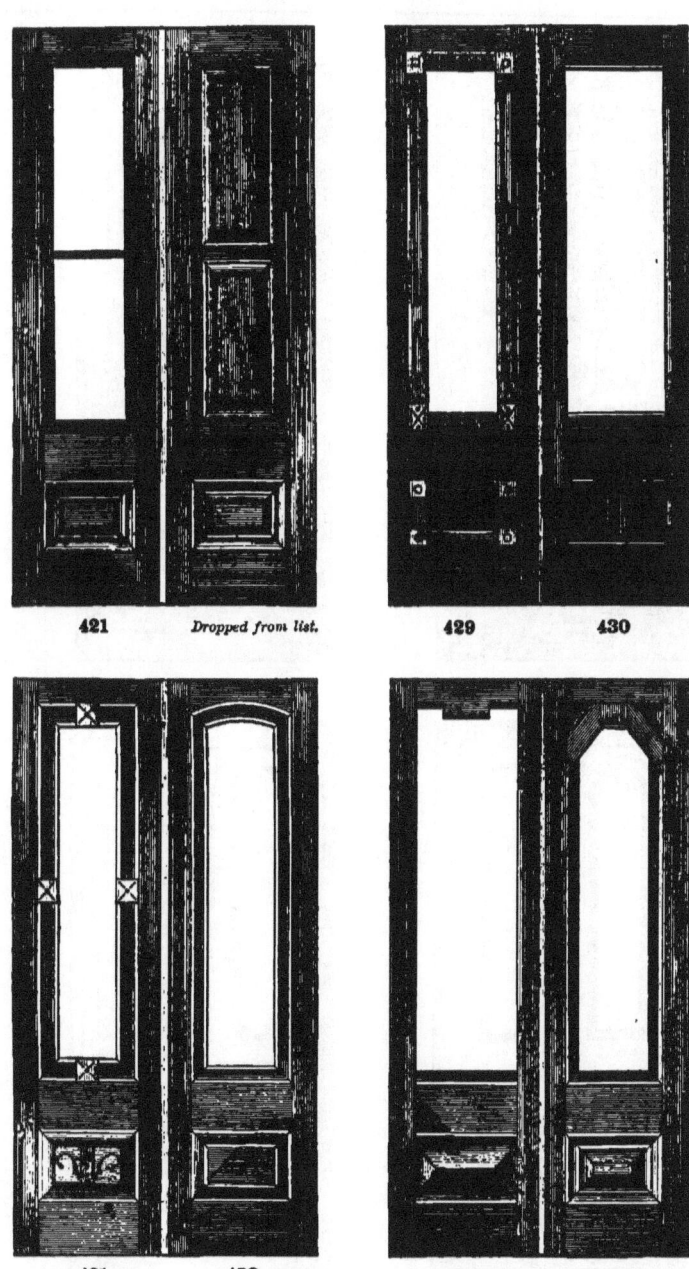

421   Dropped from list.   429   430

431   432   444   445

DOORS.

204

206

205

Dropped from list.

INTERIOR DOORS AND FINISH.

Finish 637

INTERIOR DOORS AND FINISH.

Finish 629

629

220

INTERIOR DOORS AND FINISH.

Finish 630

INTERIOR DOORS AND FINISH.

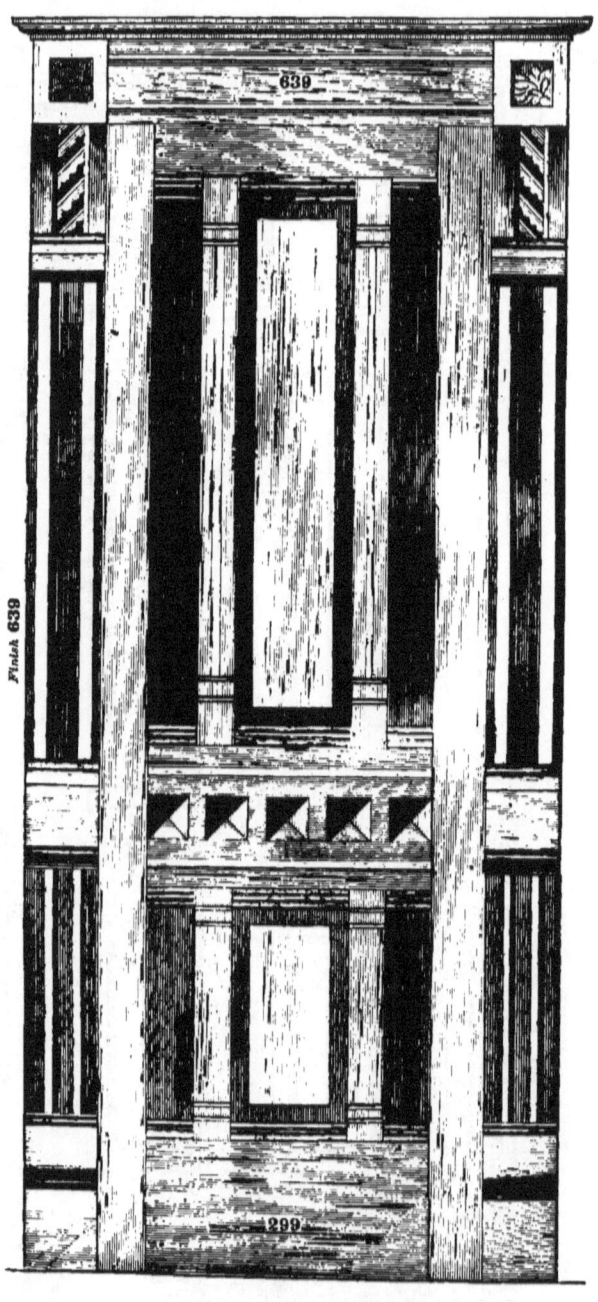

FRONT AND VESTIBULE DOORS.

329   330

FRONT AND VESTIBULE DOORS.

331   332

FRONT AND VESTIBULE DOORS.

327. 328.

FRONT AND VESTIBULE DOORS.

333   334

FRONT DOORS AND FRAME.

344   345

FRONT DOORS AND FRAME.

342   343

FRONT DOORS AND FRAME.

361

Dropped from list.

WINDOW FRAME.

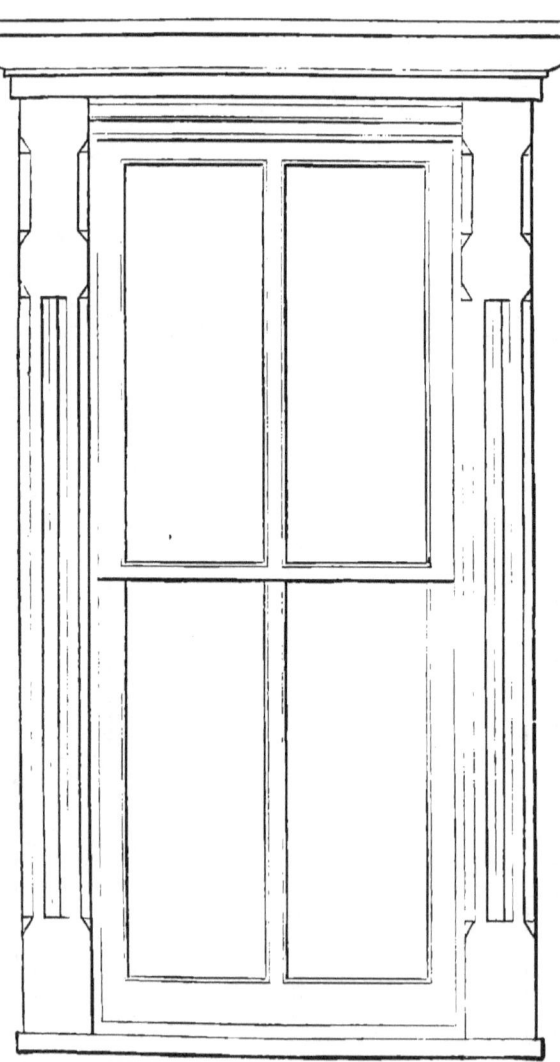

Frame No. **585**

WINDOW FRAME.

Frame No. **571**

WINDOW FRAME.

Frame No. **570**

GOTHIC WINDOW FRAMES.

Window No. 55
Frame No. 557

Window No. 57
Frame No. 558

STAIR RAILS

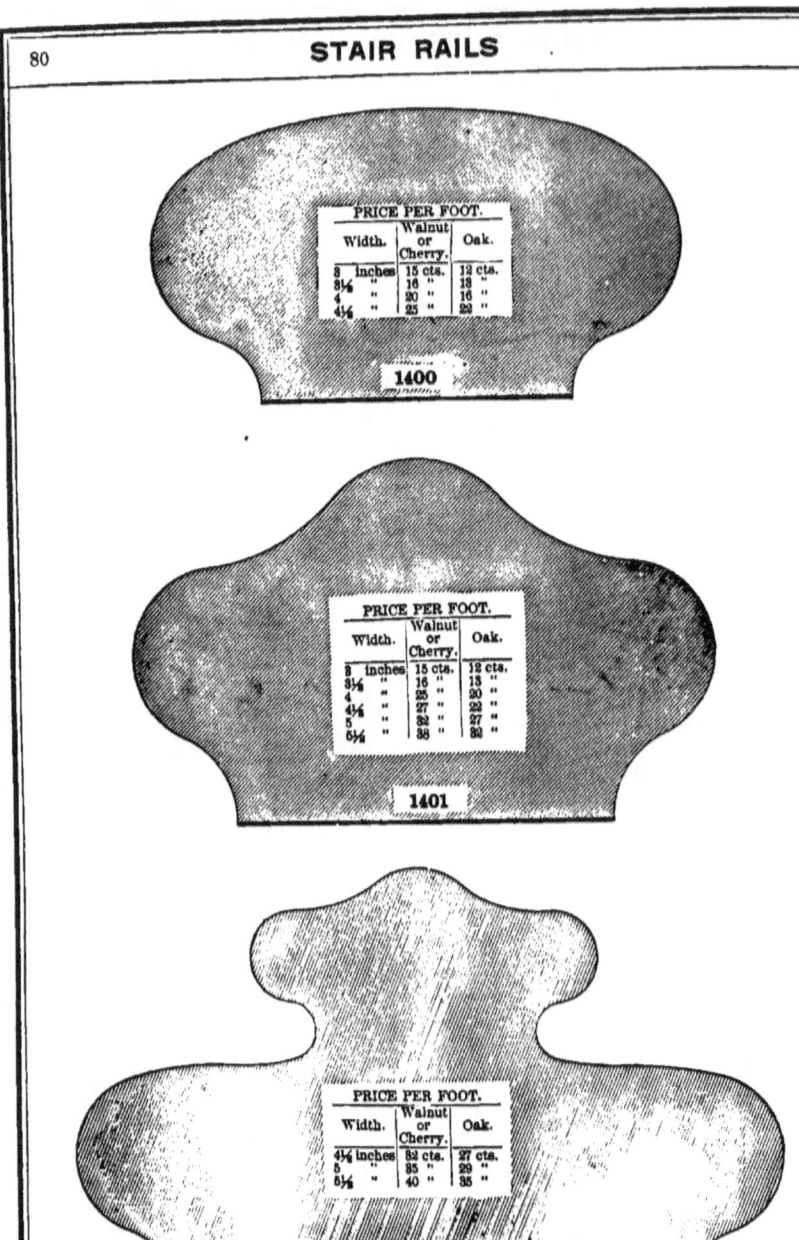

Thickness of Rails varies from 1¾ to 2¾, proportionate to width.

STAIR RAILS.

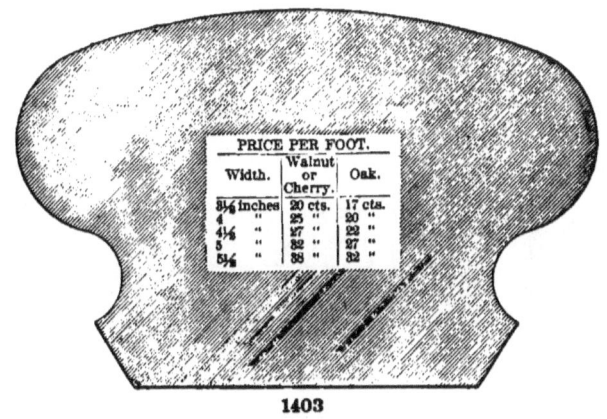

1403

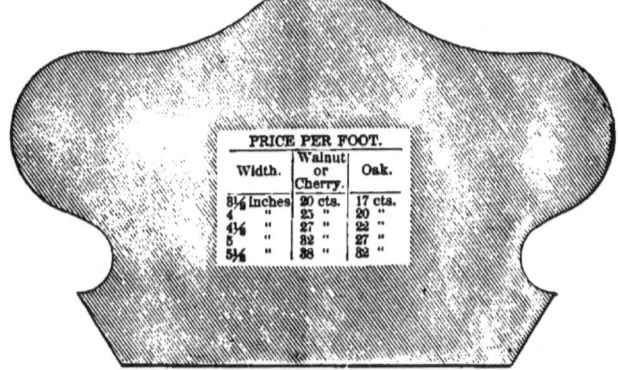

1404

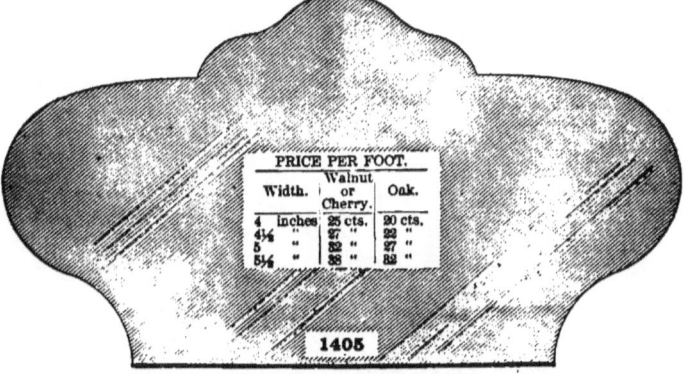

1405

Thickness of Rails varies from 1¾ to 2¾, proportionate to width.

STAIR RAILS.

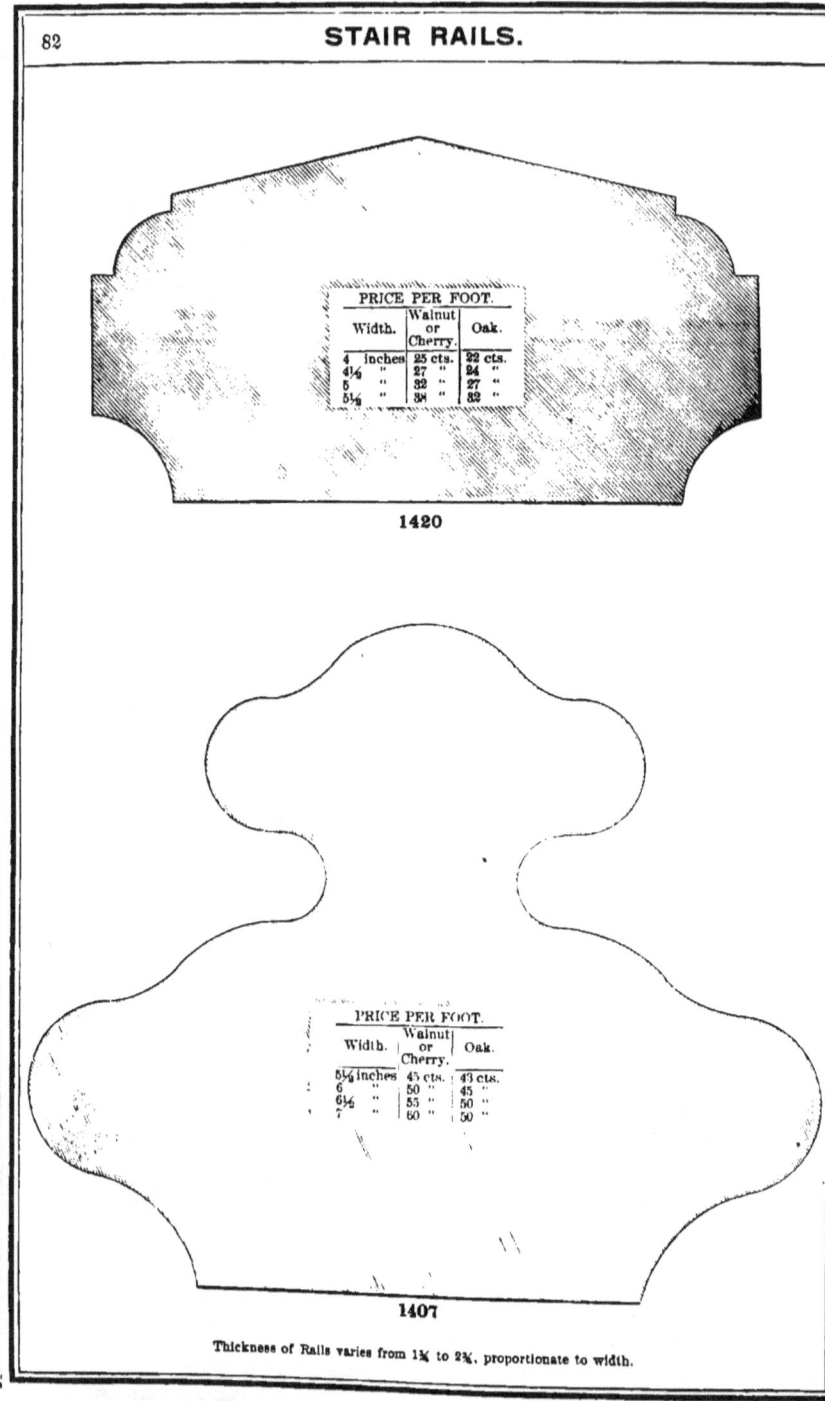

1420

PRICE PER FOOT.

Width.	Walnut or Cherry.	Oak.
4 inches	25 cts.	22 cts.
4½ "	27 "	24 "
5 "	32 "	27 "
5½ "	38 "	32 "

1407

PRICE PER FOOT.

Width.	Walnut or Cherry.	Oak.
5½ inches	45 cts.	43 cts.
6 "	50 "	45 "
6½ "	55 "	50 "
7 "	60 "	50 "

Thickness of Rails varies from 1¾ to 2¾, proportionate to width.

STAIR RAILS.

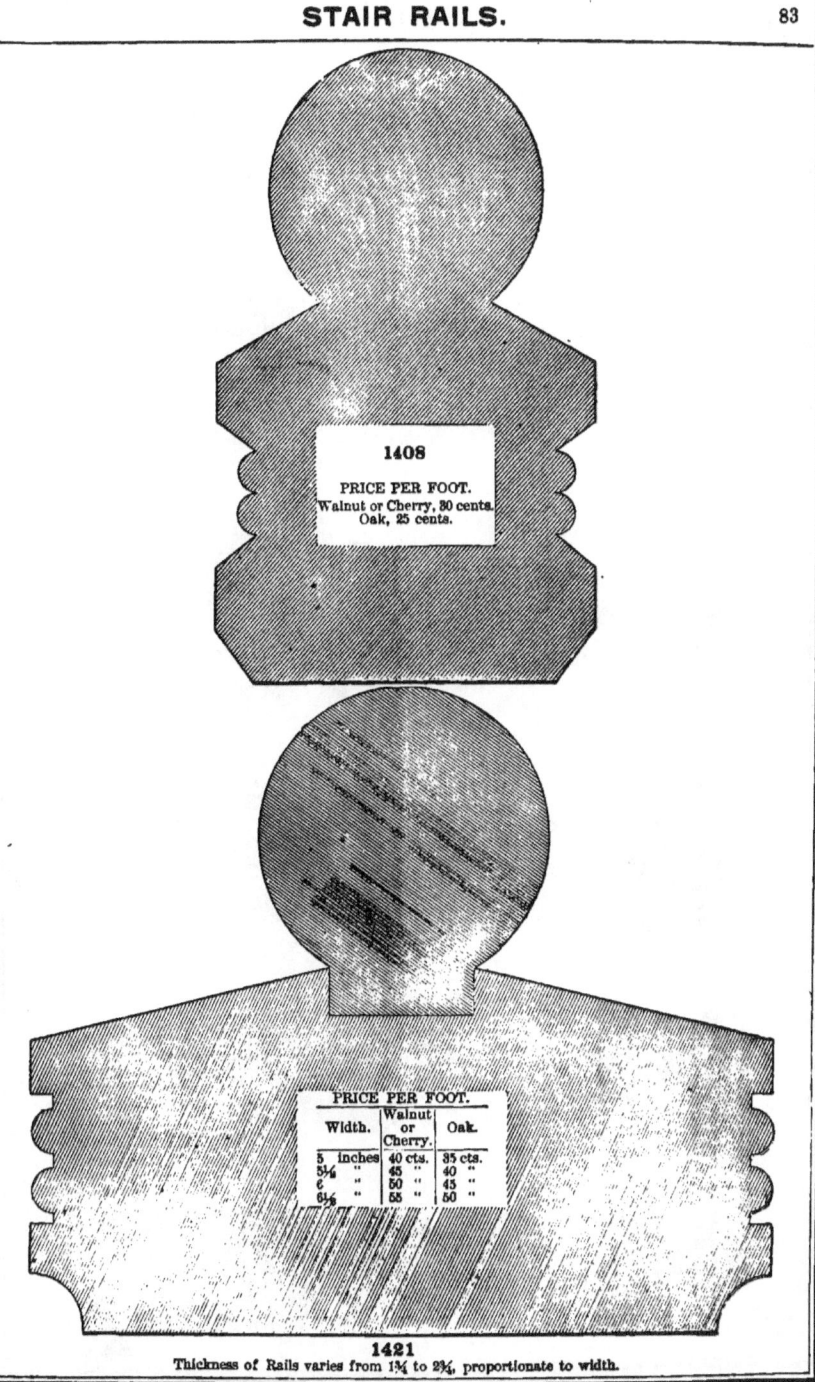

1408
PRICE PER FOOT.
Walnut or Cherry, 30 cents.
Oak, 25 cents.

PRICE PER FOOT.

Width.	Walnut or Cherry.	Oak.
5 inches	40 cts.	35 cts.
5½ "	45 "	40 "
6 "	50 "	45 "
6½ "	55 "	50 "

1421
Thickness of Rails varies from 1¾ to 2¼, proportionate to width.

NEWELS.

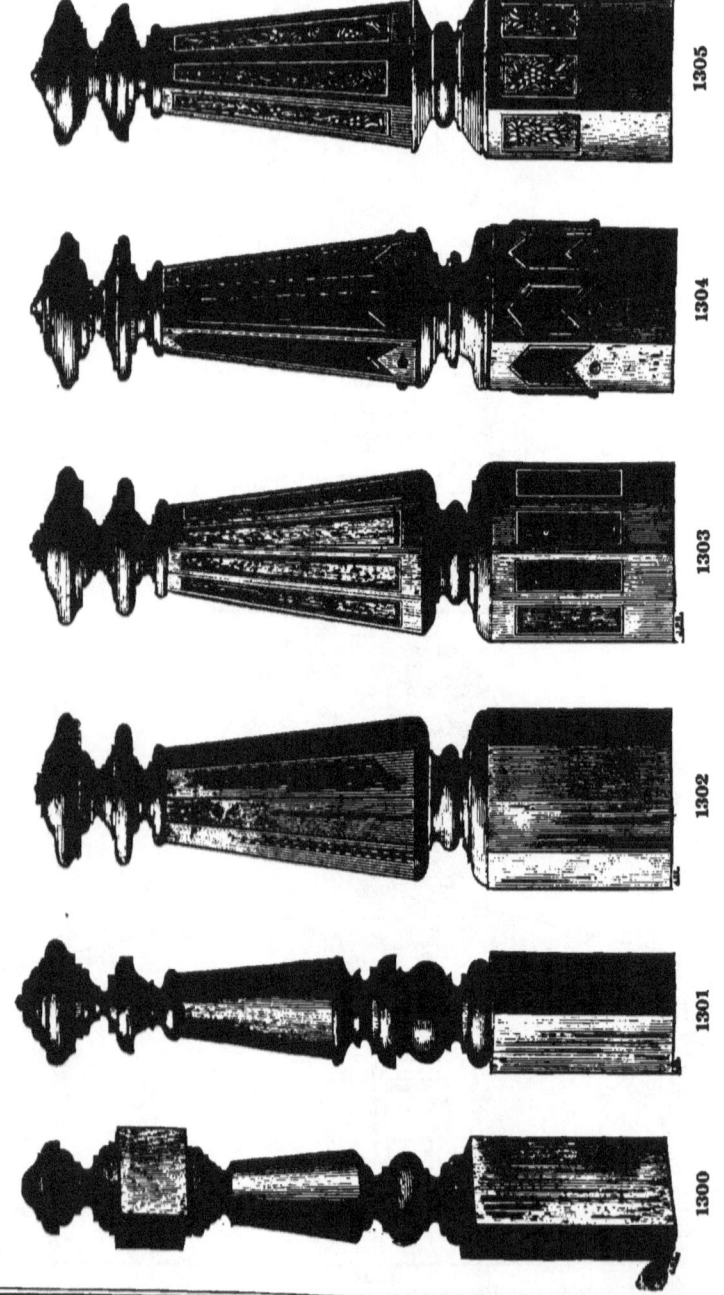

NEWELS.

1310

1352

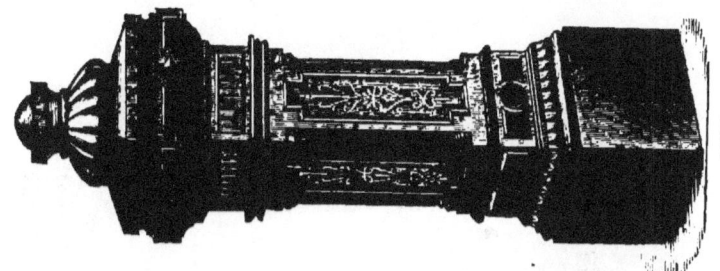

1308

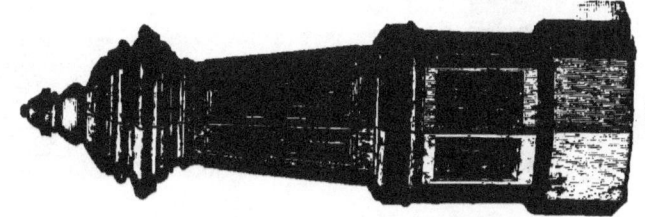

1351

1350

NEWELS.

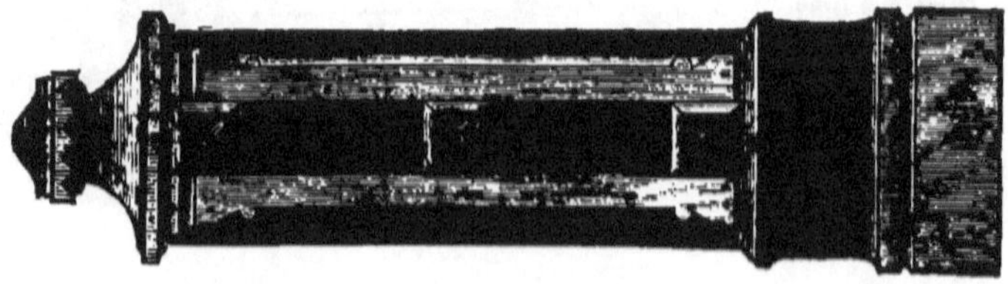

1354

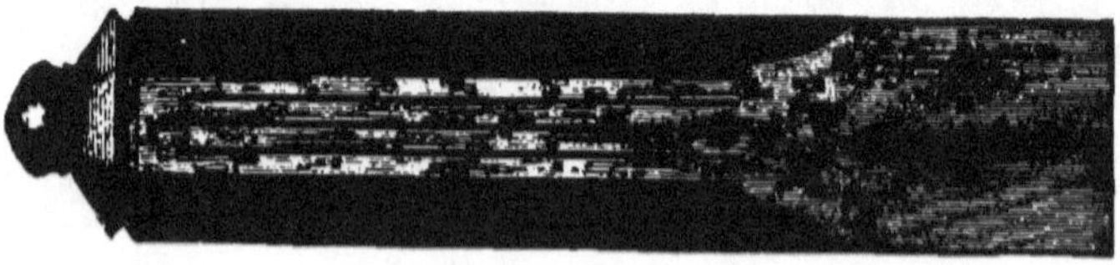

1314

1313

1353

1335

NEWELS.

1320

1319

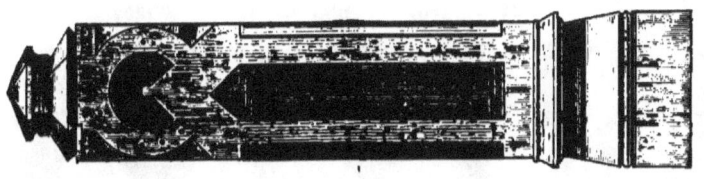

1318

1317

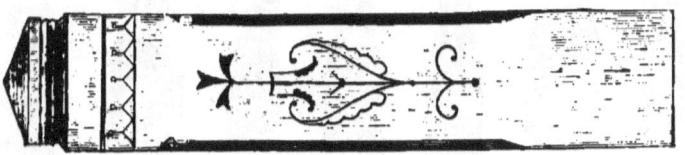

1316

BALUSTERS.

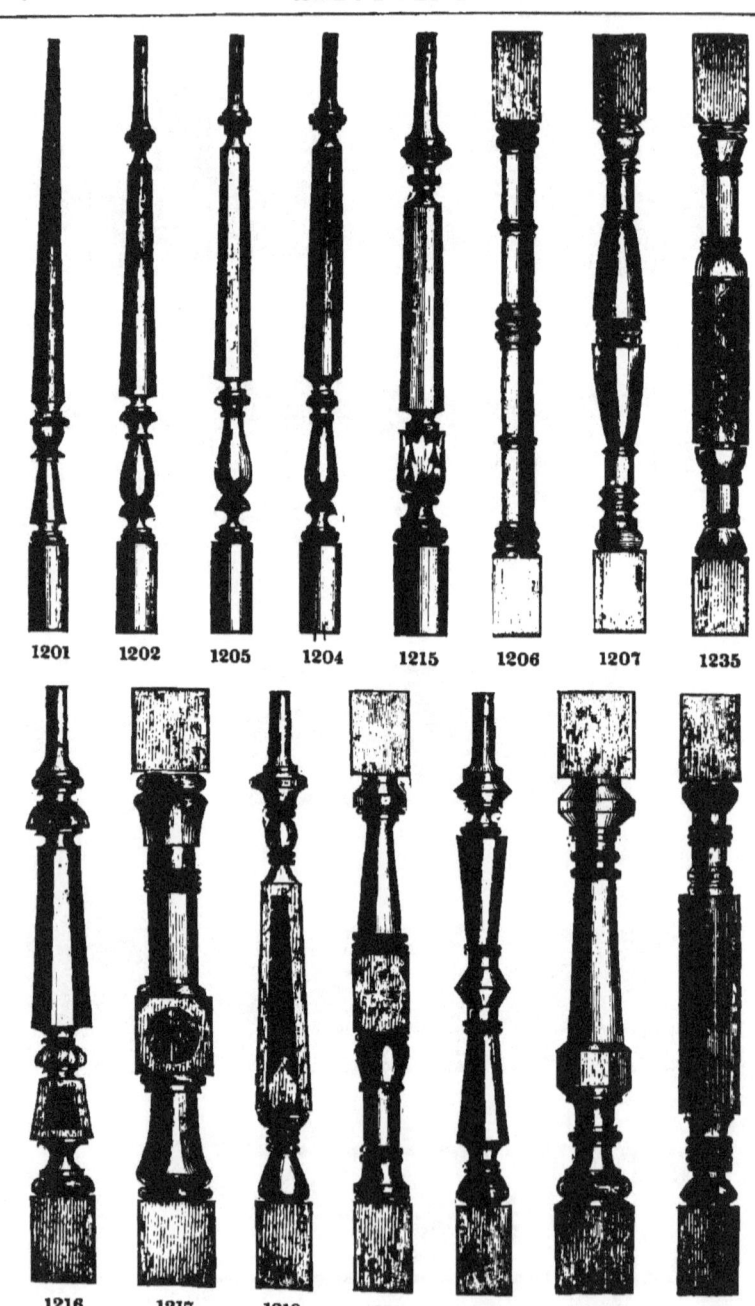

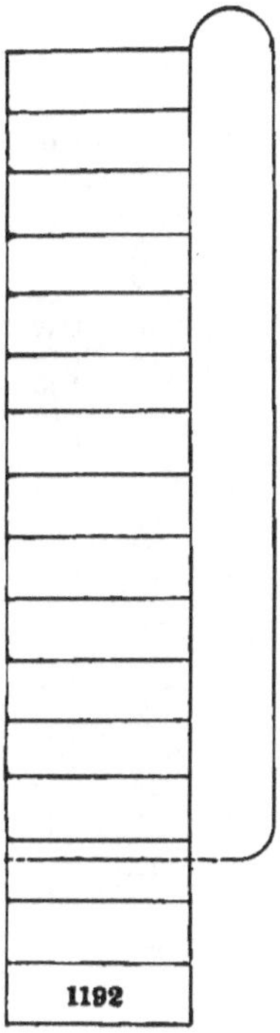

STAIRS AND STAIR BRACKETS.

1192 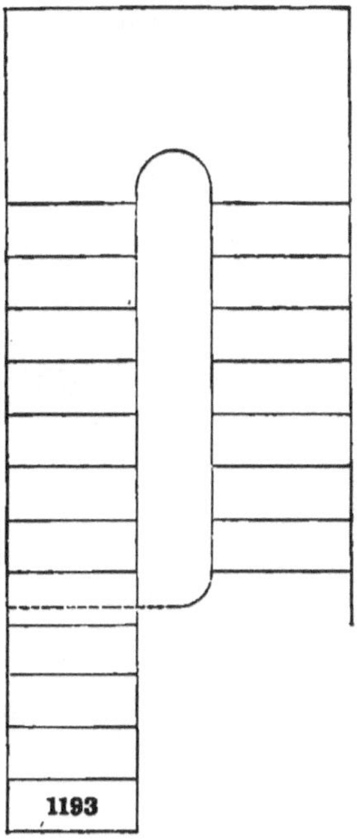 1193 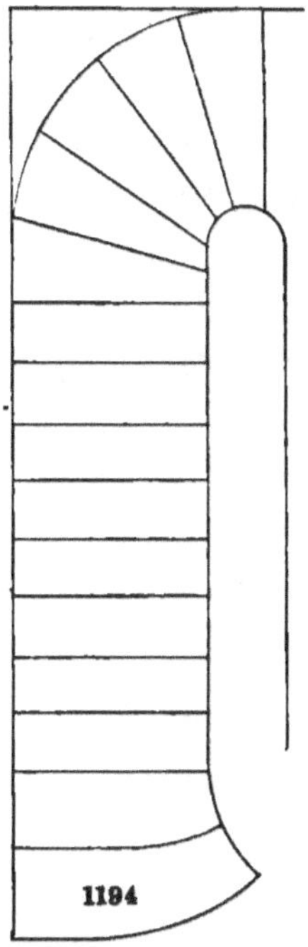 1194

1577

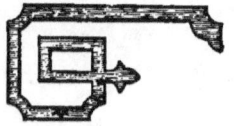

1581

Stair Brackets,
8 to 10 in. long,
Walnut 10 cts., Pine 6 cts. each.

Level Brackets for Stairs,
4 in. wide,
Price per foot, Walnut 10 cts., Pine 6 cts.

1580

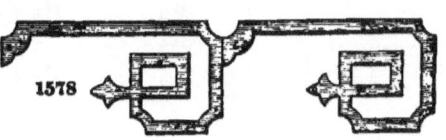

1578

PRICE OF MOULDINGS,

PER HUNDRED FEET (Lineal Measure).

Adopted January 1, 1891, by the Wholesale Sash, Door and Blind Manufacturers' Association of the Northwest.

This list supersedes all former issues in conflict with it.

No.	Price.	No.	Price.	No.	Price.	No.	Price.	No.	Price.	No.	Price.
1	$2.50	63	$1.70	125	$2.20	187	$3.75	249	$3.00	311	$1.40
2	5.00	64	2.50	126	2.00	188	5.40	250	1.75	312	2.10
3	4.50	65	4.40	127	1.80	189	4.75	251	2.00	313	2.75
4	2.00	66	16.50	128	1.60	190	3.75	252	2.25	314	4.50
5	5.50	67	11.25	129	1.25	191	1.40	253	2.50	315	5.25
6	9.00	68	6.25	130	1.10	192	1.75	254	2.75	316	3.60
7	3.00	69	4.40	131	1.10	193	2.00	255	3.00	317	3.75
8	3.50	70	1.00	132	1.25	194	3.40	256	5.65	318	3.05
9	12.60	71	1.00	133	1.60	195	5.00	257	7.45	319	4.15
10	14.40	72	4.40	134	1.80	196	8.15	258	2.50	320	2.75
11	4.00	73	1.00	135	2.00	197	5.40	259	2.25	321	2.50
12	2.50	74	1.00	136	1.00	198	4.15	260	2.25	322	2.50
13	8.25	75	1.00	137	1.00	199	3.75	261	2.00	323	2.50
14	3.50	76	1.65	138	1.50	200	2.25	262	1.75	324	1.40
15	4.50	77	2.00	139	1.90	201	2.00	263	1.40	325	1.80
16	5.00	78	2.50	140	3.40	202	1.75	264	2.25	326	2.10
17	3.00	79	4.50	141	3.75	203	1.50	265	2.65	327	2.00
18	4.00	80	3.40	142	5.40	204	1.65	266	2.50	328	2.10
19	4.00	81	1.75	143	5.00	205	3.20	267	2.75	329	2.35
20	9.00	82	1.40	144	3.40	206	2.00	268	3.15	330	2.55
21	10.50	83	1.15	145	1.75	207	2.25	269	5.65	331	2.20
22	12.60	84	1.00	146	1.00	208	3.75	270	8.10	332	2.00
23	5.00	85	1.00	147	3.40	209	5.40	271	4.90	333	2.35
24	3.00	86	1.00	148	1.75	210	1.75	272	2.50	334	2.35
25	9.00	87	1.00	149	1.50	211	3.60	273	2.15	335	2.25
26	3.50	88	1.00	150	1.25	212	2.15	274	1.90	336	2.75
27	6.75	89	1.15	151	1.00	213	2.25	275	1.50	337	2.90
28	2.50	90	1.75	152	1.00	214	5.20	276	1.40	338	1.50
29	7.50	91	2.25	153	1.00	215	3.40	277	1.55	339	2.00
30	4.00	92	.90	154	1.15	216	1.90	278	2.20	340	4.15
31	2.00	93	1.00	155	1.40	217	2.50	279	2.75	341	2.25
32	2.00	94	1.20	156	2.65	218	4.15	280	3.30	342	2.25
33	7.50	95	1.40	157	3.60	219	4.00	281	5.25	343	2.40
34	3.00	96	1.60	158	6.00	220	4.70	282	6.20	344	4.50
35	6.00	97	1.75	159	6.60	221	5.65	283	3.60	345	2.25
36	9.75	98	1.80	160	3.85	222	7.65	284	3.45	346	
37	2.50	99	1.55	161	2.80	223	4.00	285	2.90	347	
38	3.50	100	1.35	162	1.50	224	2.75	286	2.50	348	
39	4.50	101	1.10	163	1.15	225	2.75	287	2.20	349	
40	7.20	102	1.00	164	1.00	226	4.15	288	1.65	350	
41	1.15	103	2.00	165	1.00	227	2.50	289	1.40	351	
42	1.15	104	1.80	166	1.00	228	3.00	290	2.00	352	2.25
43	1.15	105	1.60	167	1.15	229	3.25	291	2.20	353	2.15
44	1.40	106	1.40	168	1.50	230	6.00	292	3.00	354	4.00
45	2.00	107	1.10	169	1.65	231	2.50	293	7.20	355	7.25
46	2.10	108	1.00	170	2.80	232	2.50	294	8.10	356	13.30
47	1.65	109	1.10	171	4.00	233	3.00	295	3.30	357	19.60
48	2.65	110	1.25	172	7.80	234	3.50	296	2.75	358	1.50
49	6.30	111	1.60	173	4.30	235	5.25	297	2.35	359	3.00
50	2.50	112	1.80	174	3.20	236	2.50	298	1.55	360	2.50
51	3.15	113	2.00	175	1.75	237	2.50	299	2.50	361	2.00
52	3.60	114	1.00	176	1.50	238	2.50	300	2.20	362	1.65
53	1.00	115	1.10	177	1.25	239	3.00	301	2.50	363	1.00
54	5.00	116	1.35	178	1.00	240	2.90	302	4.15	364	3.00
55	1.00	117	1.55	179	1.00	241	5.25	303	1.80	365	2.50
56	4.10	118	1.80	180	1.15	242	2.75	304	2.00	366	2.65
57	1.00	119	2.00	181	1.40	243	5.10	305	3.20	367	3.40
58	4.00	120	1.80	182	1.40	244	3.25	306	4.35	368	3.15
59	1.00	121	1.55	183	1.50	245	3.25	307	2.50	369	4.00
60	1.00	122	1.35	184	1.65	246	3.00	308	2.20	370	4.50
61	1.00	123	1.10	185	1.90	247	2.75	309	2.75	371	3.75
62	1.00	124	1.00	186	2.15	248	2.50	310	2.20	372	2.15

N. B.—An extra price will be charged for all Mouldings not included in the above list.

PRICE OF MOULDINGS,

PER HUNDRED FEET (*Lineal Measure*).

Adopted January 1, 1891, by the Wholesale Sash, Door and Blind Manufacturers' Association of the Northwest.

This list supersedes all former issues in conflict with it.

No.	Price.	No.	Price.	No.	Price.	No.	Price.	No.	Price.
373	$1.00	394	$1.15	415	$2.75	436		457	$5.50	478	$3.50
374	4.40	395	1.75	416	7.65	437		458	2.50	479	4.50
375	3.30	396	2.65	417	2.00	438		459	5.50	480	5.50
376	1.75	397	3.20	418	2.00	439		460	5.00	481	6.50
377	1.75	398	3.40	419	3.00	440		461	5.00	482	7.50
378	1.75	399	3.40	420	3.75	441	$6.75	462	5.00	483	3.30
379	1.75	400	1.00	421	2.40	442	3.60	463	5.00	484	7.00
380	1.75	401	1.00	422	3.20	443	6.75	464	5.50	485	7.25
381	1.75	402	1.20	423	.90	444	2.50	465	5.00	486	1.00
382	1.75	403	6.25	424	3.40	445	5.00	466	4.50	487	7.50
383	2.00	404	6.25	425	5.25	446	2.50	467	6.25	488	1.00
384	2.00	405	1.80	426	1.50	447	5.50	468	5.50	489	2.75
385	1.00	406	5.50	427	4.00	448	4.15	469	4.50	490	7.25
386	1.15	407	7.50	428	2.65	449	5.50	470	5.50	491	
387	1.00	408	3.50	429	10.00	450	4.50	471	5.50	492	
388	1.00	409	3.50	430	4.80	451	5.50	472	5.50	493	
389	1.40	410	6.40	431	4.50	452	2.75	473	5.50	494	
390	1.50	411	2.75	432	5.40	453	5.50	474	5.50	495	
391	1.65	412	8.55	433	6.30	454	2.25	475	5.50		
392	1.90	413	1.50	434	8.10	455	5.50	476	5.50		
393	1.65	414	1.75	435	9.00	456	2.25	477	5.50		

CONVENIENT FACTS FOR BUILDERS.

Bricks Required for Walls of Various Thickness.

Number for each square foot of face of wall.		Number for each square foot of face of wall.	
Thickness of Wall.	No. Bricks.	Thickness of Wall.	No. Bricks.
4 inches.	7 1-2	24 inches.	45
8 "	15	28 "	52 1-2
12 "	22 1-2	32 "	60
16 "	30	36 "	67 1-2
20 "	37 1-2	42 "	75

Cubic Yard—600 bricks in wall.
Perch (22 cubic feet)—500 bricks in wall.
To pave 1 square yard on flat requires 41 bricks.
To pave 1 square yard on edge requires 68 bricks.
One-fifth more siding and flooring is needed than the number of square feet of surface to be covered, because of the lap in the siding, and matching in the flooring.
One-thousand lath will cover 70 yards of surface, and 11 pounds of lath nails will nail them on. Eight bushels of good lime, 16 bushels of sand, and 1 bushel of hair will make good enough mortar to plaster 100 square yards.
A cord of stone, 3 bushels of lime, and a cubic yard of sand will lay 100 cubic feet of wall.
Five courses of brick will lay 1 foot in height on a chimney. Six bricks in a course will make a flue 4 inches wide and 12 inches long, and 16 bricks in a course will make a flue 8 inches wide and 16 inches long.

WEIGHT OF MOULDINGS.—1 x 1 inch, per one hundred lineal feet, fifteen pounds.

WEIGHT OF LUMBER, ETC., DRY.

FLOORING, Dressed and Matched, per 1,000 feet	1,800 pounds
POPLAR BOX BOARDS, per 1,000 feet	2,000 "
SIDING, Dressed " "	800 "
CEILING, ⅜-inch thick " "	800 "
" ½ " " "	900 "
BOARDS, Dressed one side " "	2,000 "
" and Dimension, rough, " "	2,400 "
SHINGLES per 1,000 pieces	240 "
LATH " "	500 "
PICKETS, Dressed " "	1,800 "
" Rough " "	2,400 "

PRICE LIST OF STAIR WORK.

Fancy Turned Balusters.

We turn all our Stair Balusters 2 ft. 4 in. and 2 ft. 8 in. long, and keep these lengths always in stock. Are prepared to furnish, on short notice, any length or styles desired. Odd lengths cost extra.

Prices for Fancy Turned Cherry or Black Walnut Balusters, similar to Cut No. 1201:

1½ inch Balusters	$0.10
1¾ " "	.14
2 " "	.14
2¼ " "	.18
2½ " "	.20

For Balusters like No. 1202, add 2 cents each.

Prices for Oak or Ash Balusters, similar to Cut No. 1201:

1½ inch Balusters	$0.09
1¾ " "	.12
2 " "	.12
2¼ " "	.15
2½ " "	.17

For Balusters like No. 1202, add 2 cents each.

Fluted and Octagon Balusters.

Prices for Octagon Cherry or Black Walnut Balusters, similar to Cut No. 1205:

1¾ inch Fluted or Octagon, each	$0.20
2 " " " "	.20
2¼ " " " "	.24
2½ " " " "	.26

Fluted Balusters, No. 1204, 1 cent more; Mahogany costs about double price.

Prices for Oak or Ash Balusters, similar to Cuts No. 1204 or 1205:

1¾ inch Fluted or Octagon, each	$0.19
2 " " " "	.19
2¼ " " " "	.23
2½ " " " "	.25

Fluted Balusters, No. 1204, 1 cent more.

Balusters for Outside Balustrade.

Prices for Pine or Whitewood:

3 x 3, 14 inches, each	$0.10
" 16 " "	.11
" 18 " "	.12
" 20 " "	.13
4 x 4, 14 " "	.14
" 16 " "	.15
" 18 " "	.16
" 20 " "	.17
" 22 " "	.19
" 24 " "	.21

Plain Octagon Staved Newel Posts.

Prices for Plain Octagon Staved Newel Posts, Black Walnut, Cherry, Oak or Ash, similar to Cut No. 1302:

7 inch Octagon Newel Posts, with Cap	$5.5
8 " " " " "	5.75
9 " " " " "	6.00
10 " " " " "	6.25
11 " " " " "	6.50
12 " " " " "	7.00

For Mahogany Posts, add $1.00 each; for Raised O. G. Panel, add $1.50 each.

Octagon Sunk Panel Newel Posts.

Prices for Sunk Panel Newel Posts, Fancy Moulded, Black Walnut, Cherry or Oak or Ash, similar to Cut No. 1303:

8 inch Sunk Panel Posts, with Cap	$8.50
9 " " " " "	9.00
10 " " " " "	9.50
11 " " " " "	10.00
12 " " " " "	10.50

For Circle Top Panel	add	$1.25
" Posts like No. 1305	"	4.50
" " " No. 1304	"	3.00

Fancy Turned Newel Posts.

Prices for Black Walnut or Cherry, similar to Cuts Nos. 1300 or 1301:

5 inch Newel Posts, with Cap, each	$4.00
6 " " " " " "	4.50
7 " " " " " " Walnut only	5.00

ROPE MOULDINGS.

Pine or WhiteWood	Per ft. (lineal).	Cherry or Walnut.	Per ft. (lineal).
⅜, ½ and ⅝ inch	$0.05	⅜, ½ and ⅝ inch	$0.08
1 inch	.06	1 inch	.10
1¼ "	.07	1¼ "	.12
1½ "	.08	1½ "	.14
2 "	.10	2 "	.18
2½ "	.15	2½ "	.25
3 "	.18	3 "	.30
3½ "	.25	3½ "	.35
4 "	.30	4 "	.45
5 "	.40	5 "	.60

Circles from three to four times the price of straight; cut right and left.

See section below.

www.ingramcontent.com/pod-product-compliance
Lightning Source LLC
Chambersburg PA
CBHW020154170426
43199CB00010B/1035